EARTHWORMS
For Ecology & Profit

**VOLUME II
EARTHWORMS AND THE ECOLOGY**

RONALD E. GADDIE, SR.
AND
DONALD E. DOUGLAS

COPYRIGHT © 1977 by Bookworm Publishing Company

All rights reserved. No part of this publication may be reproduced, stored in a retrieval system, or transmitted, in any form or by any means, electronic, mechanical, photocopying, recording or otherwise, without prior written permission of Bookworm Publishing Company.

Every effort has been made by the authors to insure that the information provided in this book is complete and accurate. We would appreciate any comments you have, good or bad, concerning the contents of this book.

PUBLISHED BY: Bookworm Publishing Company
P.O. Box 3037
Ontario, California, 91761

CREDITS: The authors wish to express appreciation to all those scientists and researchers who have given their kind permissions for their work to be quoted in this volume, especially C. A. Edwards & J. R. Lofty of England, S. M. J. Stockdill of New Zealand and J. E. Satchell of England. And we thank B. Lever for technical advice.

Photography—Dorian Taylor
Typesetters—Compu-Set, Pasadena, California
Printers—Delta Lithograph, Van Nuys, California

Printed in United States of America.
ISBN: 0916302-01-6
ISBN: 0916302-15-6 Hard Cover
Library of Congress Catalog Number: 76-23923
First Printing—January, 1977.

VOLUME II   EARTHWORMS AND THE ECOLOGY

INTRODUCTION

As a producer and national distributor of earthworms, I would like to share with others, the many benefits and uses of these creatures that I have learned through my world-wide studies, travels across United States and experiments on my own farm. I have talked to hundreds of farmers, gardeners and vermiculturists (worm farmers). EARTHWORMS AND THE ECOLOGY is the companion volume to SCIENTIFIC EARTHWORM FARMING which has already sold thousands of copies.

The existence and seriousness of the twin crises of pollution and starvation which worry our planet are familiar to any who has opened a newspaper, turned on a radio or television set at any time during the past decade. Indeed, the size of these problems, at times, simply seems too great for human beings to have any hope of solving them. Many observers have given themselves up to despair at the grim prospects before us. Yet, during all this time, a resource of enormous potential for alleviating these problems has been literally lying at our feet, just a few inches from the surface of the ground on which we walk. Of course, I refer to our friends, the earthworms.

The notion that these tiny creatures can be used to solve—in part—the problems both of solid waste management and of increasing the world food supply is a strange one to many, perhaps most, people in today's world. Yet the basic features of earthworm physiology and activity, which make them our potential partners in pollution control and agriculture, have been known to naturalists and scientists for nearly a century and to natural philosophers of ancient Greece for over 2,000 years.

The following chapters will explore all aspects of the earthworms, soils and crops with their interactions and interdependencies. I will examine the evidence that earthworms make a significant contribution to the fertility of soils, both in themselves and through their castings. You will learn in detail HOW earthworms may be used for soil improvement in both backyard gardens and on vast agricultural establishments.

This book explains the role of the earthworm in biodegradable solid waste management systems, from the simple Home Ecology Box to complex facilities for Annelidic consumption of municipal refuse. It will also look at the potential of earthworms as protein source for pets, livestock and even human beings. Other uses of worms, now being explored by researchers around the world, will be commented upon. Finally, I will show how and where you can begin to help earthworms help Mankind.

The California Farm Bureau Federation recently described vermiculture, worm raising, as "one of the fastest growing segments of California agriculture." That same statement is equally true of nearly twenty other states as well as of several other countries. Another observer has commented, "The earthworm today is kind of where the peanut was before George Washington Carver got his hands on it." How accurate his observation was will be easy for you to judge for yourself once you have read this book.

Ronald E. Gaddie Sr.

Ronald Gaddie Sr.
February, 1977

Ronald E. Gaddie, Sr.
President, North American Bait Farms, Inc.

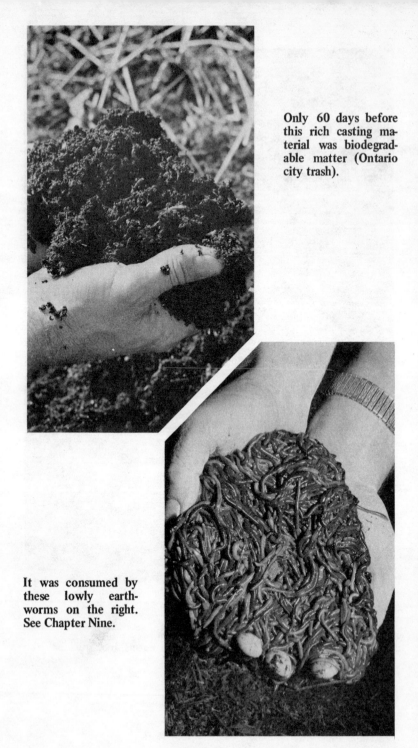

Only 60 days before this rich casting material was biodegradable matter (Ontario city trash).

It was consumed by these lowly earthworms on the right. See Chapter Nine.

CONTENTS

Introduction iii
1.) Earthworms Improve Soil 1
2.) How Earthworms Help Soils 10
3.) Earthworm Biology 19
4.) Earthworms in the Soil 51
5.) Treat Earthworms Kindly 78
6.) Your Earthworms and Farm Research 86
7.) Worms and You Together 99
8.) All About Castings 116
9.) Earthworms and Waste Disposal 127
10.) Earthworms as Food 142
11.) The Latest on Raising Earthworms Today 151
12.) Soil Types and Structures 167
13.) Plant Growth and Soil pH 185
14.) Essential Nutrients for Plant Growth 200
15.) Organic Matter in Soil 238
16.) Worms and the Future 247
17.) Books About Worms 251
Index of Major Topics 257

CHAPTER ONE

HOW EARTHWORMS HELP SOILS

For nearly 100 years, the question of what effect earthworms actually have on soil fertility has been the subject of frequent and often heated debate among agronomists, soil zoologists, biologists, entomologists, botanists, organic farmers and other farmers. Frequently, the question, "Do worms help soils?", has been lost in the heat of academic battle. Slogans and sheets covered with statistics often forgot the central question. Much of the time, this debate over the role of the earthworm has been obscured in a larger struggle between advocates of "natural" or "organic" styles of agriculture and gardening against those who maintain that through modern chemistry, Man has managed to improve on Nature's way of doing things. Until the late 1960's, the group using man-made fertilizers seemed to have the best of the argument, at least as far as can be measured from dominant practices.

Now, scientists and laymen are beginning to reach a more balanced view. Looking at the past, we can see that both sides in the *"organic/chemical"* fight had their blind spots. Advocates of chemically oriented agriculture, while amassing impressive piles of controlled studies to support their theories, have often failed to appreciate the complexity of Nature and the extent to which every process in the soil affects and is affected by every other process. "Organic" farmers usually failed to grow control fields or plots to compare to their regular methods or to use other scientific measurements which could aid their arguments.

Today we realize that all life—vegetable and animal—is bound up in a complex web whose strands cross and recross, forming a multitude of critical connections, often at unsuspected points. Just as each part of the human body is interdependent and each human being is inter-related with his society, so each plant is interdependent with its parts and inter-related with other plants, soils and atmospheres. Thus we realize that dumping chemicals from a plant in St. Louis may be killing fish near New Orleans.

On the other hand, the advocates of "organic lifestyles" have often failed to meet the established procedures of scientific inquiry. Much of the "evidence" for organic practices in the past has consisted of stories about this farmer or that gardener who achieved fantastic results from using nothing but organic fertilizers and compost, and by religiously avoiding any kind of chemical pesticide or other substance applied to soil or plant. Rarely have the organic enthusiasts subjected their beliefs to the discipline of controlled, comparative, carefully measured and recorded studies, showing the difference in results between identical plots of the same plants using two different approaches. The lack of such studies by reputable investigators, combined with the fervor which the "organic" enthusiasts often displayed, has tended to cast discredit on their theories, causing many scientists and others in positions of influence to dismiss their opinions out of hand, although we now consider those opinions were quite correct.

With this background in mind, let us examine the "hard" evidence which is available regarding earthworm effects on soil fertility. This evidence, gathered by many researchers working over many decades in widely separated areas of the world, is impressive indeed; and points almost invariably to one conclusion: Earthworms do help soils.

The following chapters will examine in detail the processes by which earthworms affect soil fertility; by reforming soil particles, increasing porosity and drainage of soils, and by the transformation of organic matter into plant nutrient materials. Each statement made in this book will be supported by one or more scientific studies, and a list of the references to these studies will be found at the end of many chapters. But the crucial yardstick for measuring earthworm effects on soil fertility, is the ultimate effect the presence or absence of earthworms has on the yields of plants. This chapter will examine such results through a brief review of these studies.

The first study examining the effect of earthworms on plant yields in this centrury was reported in 1907 by two Belgian researchers, E. Ribaudcourt and A. Combault[1], who wrote that the addition of earthworms to small test plots increased plant yields. However, as pointed out in **BIOLOGY**

OF EARTHWORMS by C. A. Edwards and J. R. Lofty, "they did not account for the effects of dead earthworms," as decayed earthworms return nitrogen to the soil. An earlier study by E. Wollny in 1890, also failed to account for all of the important variables which affect soil fertility and thus produced inconclusive results.

But in 1910, E. J. Russel reported quite clearly in the *Journal of Agricultural Science of Cambridge*[2] that he had added a half gram (live weight) of earthworms to each kilogram of soil and had obtained increased dry matter yields from plants in those soils of 25% over other soils to which dead earthworms of equal weight had been added. Then in 1922, H. G. Kahsnitz[3] claimed that the addition of large numbers of earthworms to a garden increase yields of both peas and oats grown there by 70%. Next L. Dreidax[4] told of results in 1931 of increased yields of winter wheat in plots to which worms had been added. It was the action of live earthworms that increased the soil's fertility.

In 1943, W. Haynes reported in an article appearing in *Farm Journal and Farmer's Wife*[5] on the experience of Christopher Gallup, a farmer in eastern Connecticut. According to this article, Mr. Gallup had increased yield of ear corn from 80 bushels to 196 bushels per acre by encouraging the development of native worm populations in his farm soils in just 4 years.

A scientist's count indicated that in Gallup's best fields as many as 150,000 worms inhabit each acre. Gallup built up his worm population by feeding them trash and manure. He used only a spring-tooth harrow to mix these into and to till his soil in the early spring.

Then in 1948[6] and 1949[7], Dr. Henry Hopp and C. S. Slater told of experiments they had conducted for the U.S. Dept. of Agriculture regarding earthworms and their effects on plant yields. They found that adding live earthworms at a rate of 120 worms per square meter (together with organic matter) produced a yield of 3,160 kilograms per hectare for herbage plants, but the addition of dead worms to the same types of soils produced a yield of only 280 kilograms per hectare of these plants. (A hectare is 100 meters on each side or 2.471 acres.) Further research led them to conclude that earthworms would consistently increase yields of millet, lima beans, soybeans and hay, as well as clover, grass and wheat.

The soybeans and clover had larger increases in yields than did the grass and wheat. Dr. Hopp's work in this field has been of enormous importance and caused one writer to describe him as "without a doubt the world's leading authority on earthworms."

A New Zealand scientist, R. A. S. Waters, reported in 1952[8] on an interesting experiment showing that earthworms could increase plant yields even in soils from which they had been *removed*. He first placed a quantity of *A. caliginosa* (a species of earthworm known as *L. terrestris* or common field worm in America) in a soil containing 30 grams of dry manure per kilogram of soil and allowed them to live there for 8 weeks. He then removed the worms and planted rye grass which yielded twice as much as the same grass planted in his control plots which had never had any worms added to them. This beneficial effect that worms cause will be explained in Chapter Two and later chapters.

Also from New Zealand, R. L. Nielson revealed in 1953[9] that *A. caliginosa* increased pasture production by between 28% to 100% and increased yields in mixed swards by as much as 110%.

Similar findings agreeing with those of Nielson were made by Waters in 1955[10] and also by Stockdill and Cossens in 1966[11]. Stockdill and Cossens were able to measure a pasture production increase of 72% following the introduction of the *A. caliginosa* earthworm. Their Hindon test field of upland yellow-brown earth was of interest as it was similar to many thousands of acres that had or were being developed into sown pastures.

> It was ploughed from virgin snowgrass (*Chionochloa* spp.) and fescue tussock (*Festuca novaezealandiae*) in 1940, put through a rotation of feed crops, and sown to grass in spring 1943. One ton of burnt lime was applied in 1941 and again in 1943. One ton of ground limestone was applied in 1945. It has had regularly topdressed with superphosphate and has had periodical dressings of D.D.T. and molybdenum.
>
> Earthworms were introduced in October 1949 to a relatively high producing 6 year old pasture. Four years later, in September 1953, green patches of earlier spring growth showed up at each point of introduction. (Note reversal of seasons in the Southern Hemisphere.) The 72% increase in production was measured in three cuts taken in spring 1954 from 11 year old pasture.[11]

Tests continued on this field as the scientist measured the varying yields from the areas where earthworms were fully established in the topsoil, where they were just beginning to develop their numbers and where there were still no earthworms. In 1966, this 23 year old field was producing 29% more yield where worms were established and 19% more where they were just beginning to spread when compared to the normal high yield of the heavily topdressed pasture elsewhere.

PHOTOGRAPH 1.1 – A SOIL PROFILE
This healthy soil shows the action of earthworms mixing organic material throughout the topsoil. (Photo Courtesy of S. M. J. STOCKDILL)

Also in 1953, G. Uhlen reported from Sweden that two species of earthworms had increased *barley* yields in heavily manured soils in garden frames. Unfortunately, for the American layman, his work was written in a Swedish journal which is not found in most U.S. libraries.[12]

Others from New Zealand, C. J. Hamblyn and A. R. Dingwall, stated in 1954[13] they had added lime and small colonies of only 25 worms each to sown pastures that were naturally acid. Four years later they observed that the grass around each inoculation point was both greener and denser for several feet in all directions. Stockdill reported four years later in 1958[14] that the earthworms had spread their soil enhancing activities up to *220 feet* from the original points where they had been added to the soil.

The Dutch scientist, J. A. vanRhee[15] wrote in 1965 in *Plants & Soils* that "large numbers" of earthworms added to soil *doubled* the dry matter yield of spring wheat grown there. They increased grass yields *four* times and clover yields *ten* times, although pea yields were decreased. His findings about pea crops are at variance with the earlier work of Kahsnitz, already cited, but it is possible that the two researchers were dealing with different species of peas or that there were other variable factors affecting the pea crops.

Also, vanRhee reported in 1969[16] and in 1972[17] on the results of experiments conducted on polder soils in Holland. Polders are sections of land which was originally part of the Zuider Zee in Holland. When the land was drained and put under cultivation as fruit orchards, earthworms (*A. caliginosa* and *L. rubellus*, the latter commonly known as redworms) were added to test their effects on fruit yields. He reported that trees which had about 800 worms each placed around them grew heavier root systems than did trees in soils without worms. Since root growth is an essential precondition to crop *yield,* it is presumed that the trees with worms would produce more fruit, on the average and other factors being equal, than those without earthworms.

Earthworms have also been shown to increase growth in forest trees which are important in both lumber and wood pulp production. A. E. Zrazhevskii[18], a scientist working in the Soviet Union, reported in 1957 on pot experiments in which he had increased the growth of two-year old seedlings of oak (*Quercus rober*) by 26%, and that of green ash trees (*Fraxinus pennsylvanica*) by 37% by the addition of earthworms to his tree pots. In 1972, V. G. Marshall reported to the Fourth International Congress on Soil Zoology that black spruce seedlings increased in weight "significantly" upon the addition of earthworms to the soil in which they were grown.[19]

Finally, in a book published in 1973 entitled **WHAT EVERY GARDENER SHOULD KNOW ABOUT EARTHWORMS**, Dr. Henry Hopp reported on a three stage experiment involving two different crops. His experiment is reported here in detail:[20]

> Two crops, soybeans and wheat, were planted in a silt loam topsoil that was in good structure. The soil contained a large number of earthworms to begin with and they had granulated the soil rather well. These earthworms were removed at the start of the experiment. Also the soil was fertilized heavily. Thus, the soil was put in a highly productive condition, both physically and chemically. In a second set of plantings, the good structure was destroyed by puddling and compacting the soil. A third set was likewise puddled but the earthworms were then added to restore the structure. The results of this test are given in the table below:

Soil and Earthworm Treatment	Weight of crop plant	
	Soybeans (grams)	Wheat (grams)
Natural soil structure	2.96	9.5
Soil structure destroyed	0.56	7.1
Soil structure destroyed, earthworms added	2.30	10.5

> The soybeans were severely retarded in the soil with poor structure; growth was only one-fifth as much as in the soil with good structure. Earthworm activity restored almost all the difference. The wheat did not suffer as much from poor structure; growth was three-quarters of that in the soil with good structure. Again, earthworms restored the productivity."
>
> "This experiment shows that earthworm activity becomes more important as the structure of the soil declines. Also, earthworm activity is more important with a crop, like soybeans, that is sensitive to soil structure. Most vegetables are very sensitive in this respect."

After reviewing all of these studies on the effects of earthworms on the yields of so many different kinds of plants, it is easy to understand why Drs. Stockdill and Cossens summarized their presentation to the 1966 meeting of the New Zealand Grasslands Association by saying: "The research results reported show worthwhile production increases associated with earthworm activity. Even more important is the fact that without earthworms expenditure on land development, seed, fertilizer, lime, fencing, etc. can yield only part of the true potential. Introduction of earthworms is a relatively simple and low cost operation that should be part of every land development program on soils not already populated."[21]

Nor can an objective observer disagree with the opinion of Drs. Edwards and Lofty of the Rothamsted Experimental Station in Harpendon, England, that "It is now certain that earthworms have beneficial effects on soils...." [22] Drs. Edwards and Lofty, authors of **BIOLOGY OF EARTHWORMS**, are experts in the study of earthworms. Thus controlled, scientific long term tests and experiments prove that earthworms improve plant yields. The following chapters will provide you with some background information on HOW the earthworm is so beneficial to farms, fields, gardens, orchards and forests.

★ ★ ★

REFERENCES

1. Ribaudcourt, E. and Combault, A. (1907) The role of earthworms in agriculture. *Bull. Soc. for Belg.* 212-23.
2. Russell, E. J. (1910) The effect of earthworms on soil productiveness. *J. Agric. Sci., Camb.* 2, 245-57.
3. Kahsnitz, H. G. (1922) Investigations on the influence of earthworms on soil and plant. *Bot. Arch.* 1, 315-51.
4. Dreidax, L. (1913). Investigations on the importance of earthworms for plant growth, *Arch. Pflanzenbau*, 7, 413-67.
5. Haynes, W. (1943). Earthworms: 150,000 to the Acre. *Farm J. & Farmers' Wife*; (reprinted in **HARNESSING THE EARTHWORM** by Thomas J. Barrett, 1947; available from Bookworm Publishing Co., 1976).
6. Hopp, H. and Slater, C. S. (1948). Influence of earthworms on soil productivity. *Soil Sci.* 66, 421-8.
7. Hopp, H. and Slater, C. S. (1949). The effect of earthworms on the productivity of agricultural soil. *J. Agric. Res.* 78, 325-39.

8. Waters, R. A. S. (1952). Earthworms and the fertility of pasture. *Proc. N.Z. Grassl. Ass.* 168-75.
9. Nielson, R. L. (1953). Recent research work Earthworms. *N.Z. J. Agric.* 86, 374
10. Waters, R. A. S. (1955). Numbers and weights of earthworms under a highly productive pasture. *N.Z. J. Sci. Technol.* 36, (5) 516-25.
11. Stockdill, S. M. J. and Cossens, G. G. (1966). The role of earthworms in pasture production and moisture conservation. *Proc. N.Z. Grass. Ass.* 168-83.
12. Uhlen, G. (1953). Preliminary experiments with earthworms. *Landbr. Hogsk. Inst. Jordkultur Meld.* 37, 161-83
13. Hamblyn, C. J. and Dingwall, A. R. (1954). Earthworms, *N.Z. Jl. Agric.* 71, 55-8.
14. Stockdill, S. M. J. (1959). Earthworms improve pasture growth. *N.Z. J. Agric.* 98, 227-33.
15. Rhee, J. A. van (1965). Earthworm activity and plant growth in artificial cultures. *Pl. and Soil, 22, 45-8.*
16. Rhee, J. A. van (1969). Inoculation of earthworms in a newly-drained polder. *Pedobiologia,* 9, 128-32. Also see 133-40.
17. Rhee, J. A. van (1972). Some aspects of the productivity of orchards in relation to earthworm activities. *Proc. 4th Int. Congr. Soil Zool.*
18. Zrazhevskii, A. I. (1957). *Dozhdevye chervi kak faktor plodorodiya lesnykh pochv.* Kiev. 135
19. Marshall, V. G. (1972). Effects of soil arthropods and earthworms on the growth of Black Spruce. *Proc. 4th Int. Congr. Soil Zool.*
20. Hopp, Henry (1973). **WHAT EVERY GARDENER SHOULD KNOW ABOUT EARTHWORMS,** Garden Way Pub. Co., available from Bookworm Publishing Co.
21. See reference 11.
22. Edwards, C. A. and Lofty, J. R. (1972). **BIOLOGY OF EARTHWORMS,** 168. Published by Chapman and Hall Ltd., London, England. Dist. in U.S. by Halsted Press and Bookworm Publishing Co.

CHAPTER TWO

EARTHWORMS IMPROVE THE SOIL

In the previous chapter, we reviewed 22 scientific studies and reports which concluded that earthworms increase the yields of plants. In addition to the studies cited in Chapter I, a considerable number of studies have been completed since 1972 which reinforce this conclusion. Some of these studies will be reported in Chapter 4, Earthworms in the Soil.

Just knowing that worms help plants is not enough if we are to put this fact to beneficial use in our gardens and on our farms. We must also learn *how* worms act on the soil to make it yield more fruits and flowers, food and fibres.

All plants need three elements in the soil in order to grow and produce their best: 1.) adequate moisture, 2.) room for roots and 3.) sufficient nutrients to supply their "appetite". As we shall show in the following sections, earthworms contribute to aiding or increasing the supply of all three of these needs.

A. MOISTURE

Many scientists have studied the effect of the earthworm tunnel in aiding water absorption into garden soils. R. M. Dixon and A. E. Peterson, in a paper published as part of the *Proceedings of the Soil Science Society of America* in 1971[1], suggested that water is absorbed most rapidly into the ground when the soil contains a network of channels which are open at the surface and large enough to admit "untensioned" water streams. "Untensioned" streams of water are best defined by comparing them to quantities of water moving by "capillary action."

If you fill a glass of water and then place the tip of your finger within 1/16" of the surface, almost but not quite touching, the water will actually rise up that tiny distance to wet your fingertip. This is the principle of "tensioned movement" or "capillary action" at work. Thus is water absorbed by paper towels, and by plant roots, and through moderately packed soils.

So when water falls on the ground very slowly, capillary action will be sufficient to absorb most of it. But when water comes down very rapidly, as it does during a heavy rainstorm, capillary action is much too slow; and most of the water runs along the surface of the ground without ever penetrating below. Not only is this runoff wasteful as far as benefits for the plants are concerned, but also the water may cause erosion of topsoil or surface flooding.

Dixon and Peterson, building on well established scientific principles, theorized that earthworm channels which are open to the surface would help solve this problem by allowing water to enter the subsurface region as rapidly as it fell down. When we consider the action of water flowing into a large hole in the ground, such as a gopher hole, it is obvious that the water swirls and fights against its own flow. Anyone who has tried to rid his lawn of gophers by flooding their holes with a garden hose will tell you.

Luckily, earthworm holes are much, much smaller than gopher holes. Depending on the size of the earthworm making the channel, they can vary from ¼" to 1/16" in diameter. This size is large enough, and if there are enough such earthworm channels open to the surface, they will permit water to move by gravity, flowing freely downward in an "untensioned" state.

To test their hypothesis or idea, Dixon and Peterson added *Lumbricus terrestris,* (Canadian or Native nightcrawlers) to sample lots. Then they measured the rate of water flowing into the soil with an instrument called an infiltrometer. They found that infiltration (movement of water into the soil) more than doubled following the addition of these worms, compared with the rate of movement into the same soil plots before the worms were added.

So they concluded, "Increases observed (in infiltration) are largely attributed to earthworm activity at the soil surface. Such activity not only improves the surface continuity of the channel system, but increases its subsurface continuity and depth." Yes, worms open and maintain waterways in the soil.

This study confirms earlier studies performed in New Zealand in 1969 by Stockdill & Cossens[2] which found that the infiltration rate in pasture soils was more than doubled by earthworms and the moisture holding capacity of the soil was 27% better than that of similar soil without earthworms. A later study done in Germany by Dr. W. Ehlers[3] reached the same conclusion.

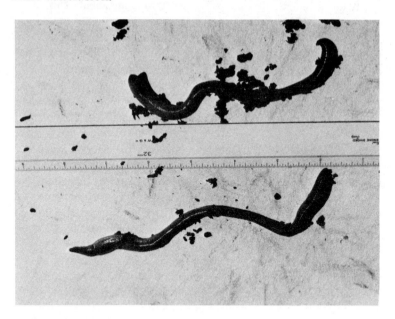

PHOTOGRAPH 2.1–L. terrestris (Canadian or Native Nightcrawler)

The basis for much of this work was laid down by Dr. Henry Hopp of the U.S. Dept. of Agriculture in the late 1940's. In one study Dr. Hopp found that the infiltration rate for water going into the soil was increased more than 4 times—from 0.2 inches of rainfall per minute to 0.9 inches per minute, after the soil has been worked by earthworms for a period of one month.[4] Surely, earthworm activity does improve the land's ability to absorb and to retain moisture.

B. AERATION AND POROSITY

The most complete account of the earthworms' effects on soil aeration is found in **WHAT EVERY GARDENER SHOULD KNOW ABOUT EARTHWORMS,** by Dr. Henry Hopp and Douglas Taff. We are reprinting their description of this process by permission of the authors.

> Earthworms are one of the most effective agents for loosening and aerating the soil. Their burrows make large passageways for the roots to grow in. They perforate the topsoil especially and gradually penetrate the subsoil, opening it for root growth

and depositing organic matter in it. But even more important is the granulation of the soil which they bring about. This is done by their production of casts from the soil and organic debris that they eat. As the soil becomes granulated with casts, it gets looser and looser. These casts are clearly visible in any soil inhabited by earthworms. During damp seasons of the year, cast production is especially prolific. At that time, casts are even deposited on the surface of the ground. However, there are always many more casts underground than there are on top.

The casts are distributed for the most part in the topsoil layer. As a matter of fact, much of the dark, granular material so characteristic of topsoil often proves, on close examination, to be earthworm casts. The dark color is due to the admixture of organic matter, or humus, with the mineral material. This mixing process takes place inside the earthworms' bodies.

The rate of cast production depends mainly on the size of the earthworm. Large species produce a greater quantity of casts than small species, and mature ones a greater quantity than young ones. An approximate rule is that earthworms produce their own weight of casts per day. This rule is based on careful measurements in controlled tests. Transposed onto a field basis, and using data from a large number of examinations in the North Atlantic and North Central states, it appears that the average quantity of soil converted into casts amounts to about 700 lbs. per acre for each day's activity. This rate of activity holds in the damp periods of the year only; for earthworms become dormant as the soil becomes dry. Assuming days of such weather, earthworms would ingest 105,000 lbs. of soil, or more than 5% of an acre plow layer per year.

It has been mistakenly assumed that the loosening of soil by earthworms has the same kind of effect on soil as cultivation with tillage implements. Some people even think there is a controversy between earthworms and the plow. This is an unfortunate misconception. They do not do the same job. From the viewpoint of simply loosening soil, tillage implements are much more effective than earthworms.

If the soil is run down, it does not stay loose after tillage. With the first rains, the clods clump together and the loose condition made by tillage is lost, Thus, tillage gives the soil only a start with good structure (besides, of course, killing the weeds), but the ability of the soil to remain loose depends upon the properties of the soil itself. This property is known as water-stability.

TABLE 5—The association of earthworms with the amount of large spaces in the soil for land in various crops, measured in the spring of the year.

Location	Condition of Land	Earthworms per Sq. Ft. (Number)	Large Spaces in the Soil (Per cent)
College Park, Md.	Bare, following corn	2	4.0
	Young wheat following corn	6	4.2
	Meadow following wheat	22	7.4
	Continuous meadow	26	12.6
Wooster, Ohio	Bare, following corn	2	1.0
	Young wheat following corn	8	2.8
	Meadow following wheat	14	4.4
	Continuous meadow	9	8.2
Holgate, Ohio	Bare, following corn	5	3.3
	Young wheat following corn	15	4.8
	Meadow following wheat	31	6.9
	Continuous meadow	38	10.7

The effect of this burrowing and granulating activity is reflected under field conditions in a close association between the numbers of earthworms and the large spaces in the soil. Examples of the association are shown in the data of Table 5. These large spaces let water and air through them readily. In size, they are spaces like one finds between the grains of a coarse sand or gravel. Meadow soils averaged 10 per cent large spaces while the more compact bare land only 3 per cent. In each location, their abundance followed the number of earthworms quite closely.[4]

As Dr. Hopp and others have shown, earthworms play a vital role in aerating the soil, and in helping the soil to maintain this good structure once it is established. It is, I think also worth repeating the opinion of another scientist that: "This one group of animals (earthworms) does more to alter the nature of the soil physically than all other soil animals combined."[5]

C. NUTRIENT TRANSFORMATION AND ADDITIONS

Many years ago it was believed that the only two elements necessary for plant growth were nitrogen and carbon. As our understanding of plants has increased, scientists have added to this list, until we now know there are, at least, 16 essential nutrients which plants must have if they are to grow well. The most commonly known, because they are the major and frequently the only components of chemically produced fertilizers are nitrogen, phosphorus, and potassium. But another three—calcium, magnesium, and sulfur—are also necessary in fairly large proportions; together with N, P & K (potassium), these make up the current list of six "macronutrients" (macro means large). Eight other nutrients are called "micronutrients" because, while they are very important, they are needed in much smaller proportions than the macronutrients. (See Chapters 12 to 15).

In Nature's plan, these nutritional elements make an endless round, being absorbed by one generation of plants, then returned to the soil in leaf fall, plant and animal decay, and finally being reabsorbed by new plant growth the following year. In all of the phases of this cycle these materials undergo continual changes, being combined with other elements in ever-differing forms and compounds.

Earthworms, where they are present or added to soils, play three highly significant *roles* in this cycle. First the worms take organic material such as leaf litter lying on the surface and pull it into their burrows beneath. Then they consume the material as food for themselves. During this process, bacteria in the earthworm gut break down much of the material into its basic elements. Finally, the earthworms excrete the material as castings and mix even more of it into fine particles spread throughout the sub-soil area.

In all of this activity, earthworms are assisted by, and in turn assist, many other soil organisms both large and microscopic. Because of the large number of different animals and organisms involved in the nutrient cycle, and the many complex ways in which they interact with each other, scientists still have much to learn about the roles played by each kind of organism, including earthworms.

However, because of their relative size, and the fact that in very fertile soils they are the most numerous of the "large" organisms, we know that earthworms play one of the dominant roles in this process. For example, it was found that in two areas studied in 1963 earthworms accounted for more of the leaf litter consumed than all of the other soil animals combined.[6] In apple orchards studied by F. Raw, who published his work in 1962, earthworms accounted for

PHOTOGRAPH 2.2—SOIL MONOLITHS
Soil structure is improved and root development is better with the addition of earthworms.

(Photo Courtesy of S. M. J. STOCKDILL)

90% of the fallen leaves removed from the surface during the winter months.[7] Drs. Edwards & Lofty, in their book, **BIOLOGY OF EARTHWORMS**, report several more studies of the same kind.[8]

Inside the earthworm, organic materials undergo a number of stages in being broken down to release their nutrient elements. First, the material is ground into very fine particles, by grains of sand or rock in the earthworm's gizzard. Scientists in Japan have found that earthworms also excrete cellulase, an enzyme which breaks down cellulose, the basic structural material of leaves and plants.[9]

Once the material has passed into the gut, it is attacked by millions of bacteria which inhabit the gut, and broken down by acids secreted by these bacteria. The bacteria in turn are assisted by a number of other enzymes secreted by the worm itself.[10]

After the material is broken down, many of the nutrient elements it contains are excreted from the worms, not only in castings, but also in secretions from the body wall; the mucous which covers every healthy earthworm.

Earthworms add nitrogen to the soil in three ways: 1.) through their castings, 2.) through urine and 3.) mucous secreted from pores along their bodies, and by the decomposition of their own bodies after they die. K. H. Lakhani and J. E. Satchell found that the worms they studied produced between 33% and 56% of their total live weight in nitrogen added to the soil each year.[11] This means that an acre of ground containing 1,000,000 earthworms would receive between 160 and 250 lbs. of nitrogen each year from those worms.

This amount of nitrogen is far more than that needed for most field crops. For example, a two ton crop of barley, removes only 72 lbs. of nitrogen from the soil. A 1½ ton crop of fine oats grown on irrigated soil will remove only 57 lbs. of nitrogen. A 2½ ton crop of wheat grown on irrigated land will need 100 lbs. of nitrogen. A 27 ton crop of sugar beets will take out 151 lbs. of nitrogen during the year, although sugar beets take most of this during a short span of time during their growth cycle.[12]

O. Atlavinyte and J. Vanagas, in a study published in 1973, found that earthworms also increased amounts of available phosphorus and potassium in the soils where they occurred.[13] Other workers have reported increases in calcium, magnesium, and molybdenum associated with earthworms.[14] (See Chapters 12 to 15.)

It is important to note that these nutrients are not simply "dumped" at one or two points on the soil surface. Rather they are produced and released gradually in small amounts spread through every layer of the topsoil. Thus, wherever the plants above spread their roots in worm populated soils, they find an abundance of the food elements they require. Earthworms, based on the evidence we have just read must surely be ranked as one of the most effective fertilizer manufacturing and delivery "systems" ever found and far more effective than any that Man himself has been able to devise. Discover your plant nutrient needs and the chemical composition of your soil, so you can use earthworms effectively to improve your crop yields.

★ ★ ★

REFERENCES

1. Dixon R. M. and Peterson, A. E. (1971). Water infiltration control: a channel system concept. *Soil Sci. Soc. Amer. Proc.* 35, 968-973.
2. Stockdill, S. M. and Cossens, G. G. Earthworms: a must for maximum production. *N. Z. J. of Ag.* 119 (1), 61 ff.
3. Ehlers, W. (1975). Observations on earthworm channels and infiltration on tilled and untilled loose soil. *Soil Sci.* 119 (3), 242-249.
4. **WHAT EVERY GARDENER SHOULD KNOW ABOUT EARTHWORMS**, op cit., p. 17 to 21.
5. In **SOIL ECOLOGY**, William A. Andrews, Ed., Prentice-Hall, 1973,pg. 56.
6. Edwards, C. A. and Heath, G. W. The role of soil animals in breakdown of leaf material. In **SOIL ORGANISMS**, J. Doeksen and van der Drift, Eds. North Holland Publishing Co., Amsterdam. pp. 76-80.
7. Raw, F. (1962). Studies of earthworm populations in orchards. I. Leaf burial in apple orchards. *Ann. appl. Biol.* 50, 389-404.
8. **BIOLOGY OF EARTHWORMS**, op. cit., pp. 141-145.
9. From *Shukan Diamond*, 8/16/75, Tokyo: Mimizu business coming out of darkness.
10. **BIOLOGY OF EARTHWORMS**, op cit. pp. 80-81.
11. Lakhani, K. H. and Satchell, J. E. (1970). Production by Lumbricus terrestris (L.). Soil biology, nitrogen. *J. Amin. Ecol.* 39 (2), 473-492.
12. United States-Canadian Tables of Feed Composition. (1969). Natl. Acad. Sci.
13. Atlavinyte. O. and Vanagas, J. (1973). Mobility of nutritive substances in relation to earthworm numbers in the soil. *Pedobiologia* 13 (5), 344-353.
14. **BIOLOGY OF EARTHWORMS**, op cit. pg. 153.

CHAPTER THREE

EARTHWORM BIOLOGY

Introduction

Why should I be interested in detailed earthworm biology? Whether you are a farmer, gardener or grower, to make the best utilization of the earthworm in your soil, you need to know how earthworms function. By knowing your earthworm's nutrient needs and its ideal environment, you can insure your earthworms will multiply quickly and produce castings at a good rate. This chapter will explain twelve topics.

1. Earthworm Classification
2. Earthworms for Commercial Production
3. "Hybrid" Earthworms
4. Earthworm Biological Functions
5. Breeding Characteristics and Life Cycle
6. Respiration
7. Digestion
8. Coelom
9. Excretion
10. pH Sensitivity
11. Moisture Sensitivity
12. Temperature Sensitivity

1. Earthworm Classification

Earthworms are invertebrate animals which form an important group of annelids, or segmented worms, found in moist terrestrial environments throughout the world. The common nightcrawler, or *Lumbricus terrestris*, is probably the best known species of earthworms. Except for size and color differences, all earthworms are physically and biologically similar to the nightcrawler. Small species of earthworms may measure one inch long, whereas giant species (like the *Megascolices australis* from Australia) have been recorded up to 4 feet in length and about one inch in diameter. Earthworms, along with leeches and marine worms, form the phylum Annelida group within the system of animal classification. The system of classification was developed by biological science to show that some type of natural relationship exists among living organisms.

Animals, insects, plants, and fishes are classified according to homology, or the fundamental identity of structure. The actual presence of homologies among animals indicates a descent from a common ancestor. The internal structure is perpetuated through long periods with little change; however, external form or appearance often changes from one generation to the other.

The system of biological classification is a very complex system with its divisions and subdivisions. There are at least seven principal divisions with the animal kingdom: Protozoa, Coelenterata, Molusca, Vermes, Arthropoda, Echinodermata, and Chordata.

This system of biological classification is described in detail in Volume 1, Chapter 2. However, a brief summary of the classification system is provided here to maintain continuity in this volume.

Within each major division are subdivisions called *phyla* (singular, phylum). A phylum is a heterogeneous assortment of organisms that comprise the animal kingdom and are based on radically different plans of organization. The members of a phylum may live in every kind of habitat, may vary in size and body form and in their method of locomotion and feeding—but they have a common basic structure.

Within each phylum, the members are further divided into groups called *classes.* Classes are generally established on the basis of a significant variation in the fundamental plan, usually in the adaptation to a special way of life.

Each class is further divided into smaller categories called *orders.* Order differences are still of such a magnitude that they can be easily recognized.

Each order consists of a number of *families* where the anatomical distinction between families is still important enough to be of survival value to the groups concerned. That is, the structures which serve as a basis of classification are probably the ones that enable a member of one family to live in a place that is uninhabitable for a member of another family.

Families are further divided into *genera* (singular, genus). The anatomical criteria used to divide a family into genera are usually so small that they are not noticed by most people and, in general, have less adaptive value for the animals concerned than they have for family distinctions.

A genus is further divided into *species*. This category is somewhat less arbitrary and represents what the biologist means as a kind of animal. There are borderline cases that do not clearly fit the species definition; however, the vast majority of animals of a species may be defined as a natural population of organisms which has a heredity distinct from that of any other group, and the members of which breed only with one another to produce fertile offspring.

In a widespread species such as earthworms, there may be a gradual change in the characteristics of the population from one end of the realm to the other so that, in some instances, widely separated individuals from the same population would be regarded as different species if there were no intergrades (intermediate or transitional forms.) When there are intergrades, such widely separated forms are regarded as belonging to different *subspecies*.

Refined observations and criteria can often distinguish among members of a species of individuals that can, but usually do not, breed or live together in the same region. They exhibit minor differences, usually more superficial than those which distinguish a species, and the groups classified on this basis are termed *varieties*, races, or strains. Still less significant are differences between the individuals of a variety or race.

The double names used to refer to specific members of the animal kingdom are called the scientific names and consist of first, the genus name (written with a capitalized initial letter) and second, the species (which is not capitalized).

1. Earthworms

Within the seven animal divisions previously stated, the earthworms in which we are interested are a part of the group *Vermes*, which consist of many different phyla. The phylum *Annelida*, to which earthworms belong, consists of four classes: *Chaetopoda, Hirudinea, Archiannelida,* and *Geophyra*.

With specific variations, the basic pattern of the phylum *Annelida* (Latin—ringed, or more often known as segmentation) suffices for about 8,500 different kinds of segmented worms as they meet the challenges of their environment. More than 5,000 species swim or burrow in marine habitats, scavenging in the bottom muds down to the deepest abysses or performing amazing mating ballets in surface waters. About 3,000 kinds of segmented worms live in fresh waters or till the soil as earthworms. A minority—less than 300 species—are predatory and parasitic leeches. Together, the phylum *Annelida* (or segmented worms) are an important part of the diet of many different aquatic and terrestrial invertebrates and vertebrates. Within the four classes of the phylum *Annelida,* the class *Chaetopoda* is the one in which the commercial earthworm grower is interested.

Chaetopoda is a Latin term which means "bristle feet." This term is descriptive of this class, since all members possess bristle-like appendages or hair-like structures, called setae, growing out of their bodies which aid them in locomotion. The class *Chaetopoda* are worldwide in distribution, living in either fresh or salt water or in moist soil. The worms of this class all breathe through their skins, although some members of the class have well-developed gills for breathing in water. The class *Chaetopoda,* which consists of over 7,000 types, is divided into two orders: *Oligochaeta* and *Polychaeta.*

The order *Oligochaeta,* of which commercially grown earthworms are a part, live mainly in fresh water or moist soil. The term *Oligochaeta* is Latin, meaning "few bristles" or, in this case "few feet." Most of them have setae, or bristles, for locomotion.

Within the order *Oligochaeta* are the various families, genera, species, subspecies, varieties, races, and strains of earthworms.

The most well-known earthworm species is probably the *Lumbricus terrestris,* or native nightcrawler. In Latin, *Lumbricus* means worm, and *terrestris* means earth. Thus, from the generic species name, our word is "earthworm."

2. Earthworms For Commercial Production

The preceding paragraphs established the fact that there are several thousand known species of earthworms in the world. However, only a few of these species are important to the commercial earthworm grower. Of these few species, only two are generally raised on a large-scale commercial basis. Some of the more common earthworm species and their relative sizes are shown in Figure 3-1.

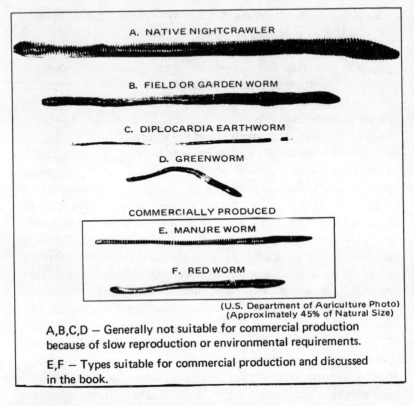

FIGURE 3.1–Six common types of earthworms and relative sizes

A. The native nightcrawler, or *Lumbricus terrestris,* is the largest earthworm in the United States and is common to the northern states. Heavy organic fertilization of the land seems to increase nightcrawler population in meadows, fields, and lawns. It is a slow reproducer, does not survive well in shipping, and is not raised on a large-scale commercial basis because of the slow production rate and required specialized controls. Commercially raised nightcrawlers are generally smaller than their native counterparts.

B. The common field worm, or *Helodrilus caliginosus* (also classified by some as *Allolobophora caliginosa*), occurs throughout the humid areas of the country. It is more common than the native nightcrawler, particularly in the southern states. The field worm may prevail in the same locality as the nightcrawler if the fertility level is too low for the nightcrawler. The field worm is not a good reproducer nor particularly adaptable to commercial production.

C. The *Diplocardia verrucosa* is a small, slim earthworm which has no English name. It is found in many moderate-to-colder climate areas in soils of low fertility. The soil may contain a large number of these worms. However, they have very little effect on the soil since their burrows and casts are so small. This worm is not adaptable for commercial production.

D. The green worm, or *Helodrilus chloroticus,* is a short, stout, greenish-colored worm. It is a very inactive worm and is more often than not found in a semidormant state, while other worms are active. This worm is not adaptable to commercial production.

E. The manure worm, or *Helodrilus foetidus* (also classified by some as *Eisenia foetida*), is known as the bandling or red wiggler because of its squirming reactions when handled. This species can be distinguished by the transverse rings of yellow and maroon which alternate the length of its body. It is normally found in manure and refuse piles. This worm is particularly adaptable for commercial production.

F. The red worm, or *Lumbricus rubellus,* is also basically a manure worm. However, the red worm differs from the manure worm (E) in that it is a deep maroon color but does not have the yellow stripes of the manure worm. The red worm is also normally found in manure and refuse piles, sometimes along with the manure worm. It is very adaptable to commercial production, as is the manure worm.

Neither the manure worm (E) nor red worm (F) is common to agricultural land; however, they will come in where large amounts of refuse or organic matter have been added to the soil. This worm may also be added to the soil for soil improvement, provided sufficient organic matter is available for food.

Lesser Known Commercial Types

There are two other types of earthworms (not illustrated) which are also raised commercially: the brown-nose angleworm and the African nightcrawler.

G. The brown-nose angleworm, or *Allolobophora (Eisenia) foetida*, is light reddish-brown in color, thicker bodies, and about one-third larger than the red worm. The brown-nose angleworm may show traces of the characteristic stripe of the red wiggler. This worm also thrives in the soil when organic materials are available as food. Although it is a good breeder and shipper, it has not obtained a great popularity within the industry.

H. The African nightcrawler, or *Eudrilus eugenie,* is almost as large as the native nightcrawler. It is harder and more costly to raise than the manure and red worms, but commands a premium bait price because of its size. Africans must be raised under strict environmental controls, and for that reason are not considered for general production in earthworm farming or for use in soil improvement.

Although other types of worms are discussed in this chapter, this book is primarily concerned with only the manure worm (E) and the red worm (F), hereafter referred to as earthworms. It has been estimated that these earthworms, commonly found together, comprise 80 to 90 percent of commercially produced worms.

These are the only two types of earthworms which have consistently been "domesticated" (i.e., raised in captivity and specially fed and watered for increased size and reproduction). These earthworms have adapted themselves to living in many different types of environment and temperature and to eat whatever food is available. They will eat almost any organic matter at some stage of decomposition, as well as many other types of materials, as long as it can be ingested and digested.

Domesticated earthworms are larger than their native counterparts, but retain the native toughness, liveliness, attractive coloring, and prolific breeding properties. The domesticated earthworm is so much easier to raise than any other type of earthworm and the expense of raising it is so much less that it is greatly preferred by earthworm growers. Detailed information on the commercial production of these earthworms is provided in Volume 1 of this book.

3. "Hybrid" Earthworms and Confusing Trade Names.

The term "hybrid" earthworm is widely used by commercial earthworm growers across the country. However, despite the many conflicting views and theories of writers, researchers, and growers, it is highly doubtful that there is such a thing as a *hybrid* earthworm.

Earthworms are so adaptable to environment that if proper food and moisture are available, they will grow to resemble, to some extent, any other type of earthworm with which they are forced to live. Experiments have shown that almost all of the manure worms, if mature, will grow up to 5 to 6 inches long within a period of 2 to 3 weeks when forced to live with the African nightcrawler. This is sometimes called homomorphous development and is common in many forms of life. It allows earthworms to grow as large as necessary in order to compete for food or to survive in a given environment. It is almost impossible to establish one species of common earthworm and to maintain it without other types showing up. Wild garden worms have a habit of finding their way into places with proper environments, and the earthworm grower expends much effort to provide such an environment. These facts lead many growers to claim they have developed a hybrid or cross, and they may actually think their claims are valid. Sometimes dissection and microscopic examination may be required to identify the homomorphously developed earthworm to the correct species.

The species (G) *Allolobophora (Eisenia) foetida,* often known as the striped worm, tiger worm, or California Striper, is one of the few earthworms known to be a cross between two species. However, the cross was not accomplished by man but by nature over a period of unknown time.

Differences in environment, feed, etc., may produce temporary changes in size, color, and general appearance. This temporary change, as well as good salesmanship on the part of individual growers, has resulted in a variety of names used by growers in describing their earthworms. Some of these are Red Hybrids, Red Wigglers, Egyptian Reds, California Reds, California Striper, Red Gold Hybrids, and many, many more.

These many names all describe the same basic earthworm species—either the manure worm, the red worm, or both. Some domesticated earthworms are better than others; but that is due, in almost all cases, to better feeding, handling, and care. The names and the fact that they are, or are not, hybrid is not really important. The important fact is that earthworms are a valuable asset to the soil and, therefore, to man.

4. Earthworm Biological Functions

The earthworms' biological metabolisms are highly important to the function and survival of the earthworm in the soil. It is these metabolic functions which allow the earthworms to convert dead and decaying organic matter into the finest of topsoils at a rate faster than nature and with a mineral content higher than that produced through natural decomposition of the same organic matter. No one has yet fully explained how the lowly earthworm is able to accomplish these phenomenal feats even though exhaustive studies of the earthworm have been made over a period of years.

Following is a simplified description of the overall physical biology of the earthworm to familiarize the reader with the earthworms' biological structure. A detailed discussion of the physical biology of the earthworm is provided in Volume 1, Chapter 2.

The biological structure of the earthworm varies slightly among all earthworm species. The various species, as already noted, are so closely related that even the experts are sometimes confused in their attempts to separate and classify the various species.

The native nightcrawler, *L. terrestris,* is probably the best known earthworm species, and is commonly studied as representative of the earthworm because of its large size. The nightcrawler is large enough for dissection, is found in most parts of the world, and varies little in size and color. The biological structure of the nightcrawler is physically and

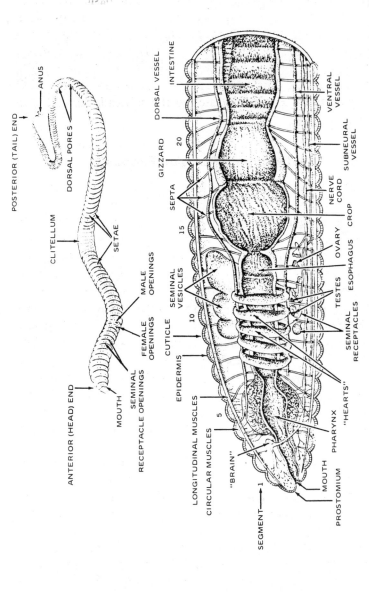

FIGURE 3.2—Major parts of earthworm (L. terrestris) reproductive, circulatory, nervous, digestive, excretory and muscular systems.

biologically similar enough to all other worms commonly found in the earth that the slight differences make no difference to the earthworm grower. Therefore, the following simplified discussion is based on the biological and physical structure of the nightcrawler.

If there were a scale of increasing animal complexity from the lowly protozoa to insects or invertebrates, the earthworm would occupy the top position on the scale. The earthworm has one of the more highly developed nervous, closed-loop circulatory, digestive, excretory, muscular, and productive systems of any animal within its group. The major elements of the systems are shown in Figure 3.2. Figure 2-2, Volume 1, is a cross-sectional view which locates all of the earthworm organs.

Segmentation

Eons ago earthworms originated from the aquatic protozoa, and in some systems retain certain aquatic characteristics. The most noticcable feature of the earthworm is the ringing of the body, which is not merely external but involves nearly all of the internal structures. The name of the phylum to which the earthworm, leeches, and marine worms belong is *Annelida* (Latin, ringed). The ringed condition is more often known as segmentation, and each ring is called a segment. These segments play an important part in the life of the earthworm and in the work of those interested in breeding them commercially.

The nightcrawler (*L. terrestris*) can be distinguished from the commercially grown manure and red worms, *Helodrilus foetidus* (*Eisenia foetida*) and *Lumbricus rubellus,* by the number of segments. The nightcrawler has approximately 150 segments, while the manure and red worms have only approximately 95 segments.

Many theories have been advanced to explain the origin and function of segmentation. However, segmentation in one phylum of animals is not necessarily the same as in another. Therefore, one explanation cannot be devised to cover all events of segmentation, and it is impossible to pinpoint the common ancestor of all segmented animals, since segmentation probably arose independently in more than one line of evolution.

Segmentation within the earthworms seems to have the same general advantages as the dividing of the animal body into cells. That is, it is possible for each segment to specialize in different functions.

Anterior or Head

The first part of the earthworm, the anterior end or head, is not a segment; it is the prostomium, which serves as a wedge to force open a crack into which the earthworm may crawl and as a covering for the mouth. Chitinous hair-like structures, called setae (bristles), are located on each segment. (Chitin is the substance which makes up the hard outer covering of insects.) The lack of protruding structures, other than the setae, on the earthworm's body is an adaption that permits efficient burrowing. In addition, various epithelial (or skin) glands secrete a lubricating mucus that facilitates movement through the earth. A thin, protective cuticle is also formed by the epithelium.

Body

The body wall of earthworms is heavily muscularized. The muscles are arranged within the wall in a circular as well as in a longitudinal fashion. The fluid-filled cavity, the coelom, separates the body wall and the digestive tract. In earthworms, the coelom is a true coelom because it has a cellular lining. This true coelom sets the earthworm apart from the lower worms, which either lack a body cavity or have a pseudocoelom.

Within the earthworm a septum, or partition, separates each segment internally. However, small perforations in the septa allow movement of fluids from segment to segment. This fluid movement is important in the circulation of nutrients and wastes within the body.

Blood Circulation

In addition to circulation through the septa, the earthworm has a highly developed closed-loop circulatory system. A dorsal vessel above the intestine carries blood forward, and a vessel below the intestine moves blood toward the rear. These two major vessels are connected by smaller vessels encircling the intestine. Other vessels carry blood to the excretory organs and gonads (sexual organs) located in the individual segments.

Digestion

The earthworm's digestive tract is highly adapted to its burrowing and feeding activities. As the earthworm moves through the soil, portions of the soil are swallowed; the organic detritus (debris) and small soil organisms are separated from the organic soil particles and are then digested. Earthworms have evolved highly muscular digestive tracts which allow the movement of large volumes of coarse materials through the intestine. The fore part of the intestine has a heavy crop and gizzard which also produce mucus and fluids.

Earthworms lack specialized gill surfaces or breathing devices. Respiratory exchange, or ventilation, in the earthworm occurs across the body surface.

Reproduction

The earthworm's reproductive system is quite complex. Earthworms are hermaphroditic—that is, each individual possesses both male (testes) and female (ovaries) reproductive organs. A *mutual* exchange of sperm occurs during copulation, and the sperm is temporarily stored in cavities called seminal receptacles. The sperm and egg cells are released and mature in the coelom, where the sperm cells enter special pouches called seminal vesicles, or sperm sacs. Similarly formed pouches, called ovisacs, are where the egg cells mature. Mature sperm and egg cells are conveyed to body openings (genital pores) by a duct sytem. In sexually mature earthworms, the body wall of the forward segments is thickened by gland cells, forming a more or less conspicuous girdle known as the clitellum. The clitellum secretes band-like cocoons in which the sperm and egg cells are placed. These cocoons are then deposited in the soil until the small earthworms hatch.

Senses

Numerous light-sensitive cells are present on earthworms, particularly on the anterior (head end) surfaces. These cells enable earthworms to detect, and thus avoid, harmful light conditions. Other areas of the body surfaces are sensitive to touch and vibrations, enabling earthworms to rapidly withdraw into their burrows when certain stimuli are received.

The earthworm has a brain located in the head, or prostomium. However, removal of the brain does not seem to seriously affect the earthworm's ability to move, burrow, or feed.

Earthworms are usually most active in the spring and fall in temperate regions, hibernating in the winter and estivating in the summer.

5. Breeding Characteristics and Life Cycle.

In their natural habitat, earthworms follow a well-defined yearly cycle. This cycle might be considered as starting in the fall of the year as shown in Figure 3.3. At that time many of the earthworms are young and when wet, cool weather occurs, they become extremely active physically. This high level of physical activity normally continues throughout the fall, winter, and spring. During the winter, both mature and young earthworms can be found in the soil, and by late spring most of the earthworms are mature. When summer arrives, the soil heats and dries and the earthworms become less and less active. They lay eggs, and may die from the heat and dryness. During the hottest and driest part of the summer, almost all of the earthworms in the soil are young or unhatched eggs that were deposited by the mature earthworms before they died. Thus, summer is a period of sharp decline in physical earthworm activity, and is mainly a period during which the generation starts for the next year.

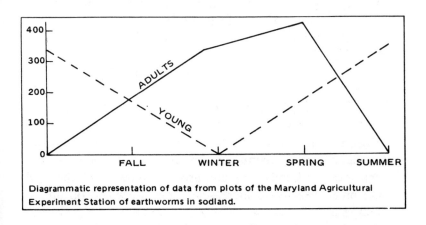

Diagrammatic representation of data from plots of the Maryland Agricultural Experiment Station of earthworms in sodland.

FIGURE 3.3—Annual cycle of earthworms in natural habitat

This cycle of the earthworm in its natural habitat is a reflection of seasonal changes in the weather. Differences in weather from year to year, or from one region to another, can somewhat modify the cycle. It can also be modified by keeping the soil moist and cool during the summer months through watering or mulching; the earthworms will then be physically active throughout the year, with exception of the coldest winters.

Earthworm egg capsule production is higher in the spring than any other season of the year. Breeding occurs frequently in the spring, possibly every two weeks, slowing down during the summer and becoming infrequent in the winter. While, as stated, under correct environmental conditions, some earthworms will produce capsules all year long, seasonal changes in the soil environment will cause a variance in the number of capsules produced by different species. Therefore, statistical data for one species can be different from that for another species.

Earthworms Are Not Natural Breeders

They will breed only:

a. To propagate the species if faced with destruction by adverse environmental conditions. Since earthworms were originally water animals, they require plenty of moist soil. Thus, if a drought occurs and the soil starts to dry, earthworms will breed heavily and lay capsules. During the drought, the earthworms themselves will die and the capsules will lie dormant. When rain occurs and the soil becomes moist, the capsules will hatch in a few days. Earthworms also seem to sense freezing weather, and since any earthworm will die if exposed to freezing temperatures, they will again breed heavily and lay capsules to propagate the species.

b. To increase their number under favorable environmental conditions. If earthworms are placed in a favorable environment—one with correct moisture, plenty of food, loose soil, and proper temperature—they will breed heavily to increase their number and fill up the space.

c. If force-bred by placing a large number of mature breeders in a given area. Overpopulation of a given area places the earthworms closer to one another so they do not have to search for a mate.

Ideal Environments

The following are environmental conditions which affect the rate of earthworm breeding, capsule production, and hatching of young earthworms.

a. Temperature

An earthworm will die if exposed to freezing or lower temperature, but will live and continue to breed in heat up to 100 °F or higher in well-shaded locations if plenty of moisture is always present. Drying out of soil quickly affects earthworms and may stop reproduction. Earthworms are most active when at temperatures of from 60 °F to 80 °F. Intensive capsule production will occur between 60 °F and 70 °F, provided sufficient moisture and feed is available.

b. Moisture

Earthworms require plenty of moisture for growth and survival. The soil should be crumbly moist, not soggy wet. A soggy wet soil can create an anaerobic condition and a subsequent build-up of carbon dioxide which will drive the earthworms from the soil. An earthworm will also produce fewer capsules when the soil is either too dry or too wet.

c. Feed

An adequate amount of organic material in the form of animal manure or green organic matter should be available at all times. The organic material should be moist for easy assimilation by the earthworm. An earthworm will produce more young per capsule if fed a feed which is high in nitrogen and vital nutrient minerals. The earthworm needs protein, which is necessary for cell generation and growth, and protein is a constant component of nitrogen (i.e., the protein content of a feed is 6.25 times the nitrogen content). Edwards and Lofty[1] cite a study by Guild and Evans (1948) whereby two species of earthworms which were fed bullock and sheep droppings produced 6.4 and 6.7 times more capsules, respectively, than when fed oat straw. The same two species were then again fed the bullock and sheep droppings and then fodder. The results were even more dramatic; those fed the animal droppings produced 8.4 and 8.8 times more capsules, respectively, than those fed fodder. This study seems to be conclusive proof of the statement in Volume 1 that earthworms fed a nitrogen-rich animal manure diet will grow faster and produce more capsules than those fed other materials.

d. Stimulants

There are no known stimulants which will force earthworm breeding. Fairly fresh manure or other nitrogen-rich green organic matter seems to be the best stimulant to rapid breeding.

Under ideal conditions, the breeding and growth cycle for commercially raised earthworms is as follows (see Fig. 3.5):

a. A mature earthworm breeder will produce an egg capsule every 7 to 10 days, or 36 to 52 capsules per year.

b. The egg capsule will hatch in approximately 14 to 21 days. Each egg capsule will hatch from 2 to 20 young earthworms with an estimated average of 7.

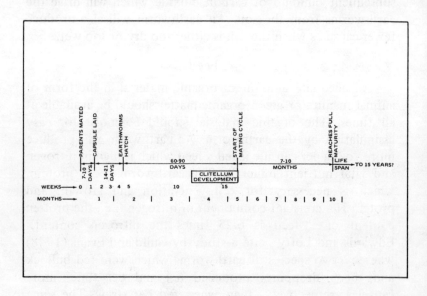

FIGURE 3.4—Earthworm life cycle under ideal conditions

c. The young earthworms will mature to breeding age in approximately 60 to 90 days with proper feed and care. This does not mean that the earthworms are full grown. It generally requires several months to a year for an earthworm to reach full mature size, averaging 3 to 4 inches in length.

It is possible for one mature breeder to produce a conservative 1,200 to 1,500 offspring in a year under favorable year-round moisture, food, and temperature conditions. Even under average natural environmental conditions, it is estimated that 1,000 mature breeders can produce more than 1,000,000 earthworms and capsules in a year or close to 1,000,000,000 earthworms and capsules in two years.

The preceding observations were originally made by one of the pioneers in the commercial earthworm field, Dr. Thomas J. Barrett,[2] in the early 1940's and similar results have since been observed by commercial growers. Scientific studies by Evans and Build (1948) and Wilcke (1952)[3] are summarized in Table 3.5. While this data shows somewhat more extended periods, it is under average conditions while the preceding data is under ideal breeding and development conditions and, as stated, has been reported by many commercial growers. Michon (1954)[4] indicated clitellum

TABLE 3.5–Time of development for manure and red worms

Species	No. of Capsules per Worm per Year	Incubation Time of Capsule (weeks)	Period of Growth of Worm (weeks)	Total Time for Development (weeks)
L. rubellus (Red Worm)	106	16	37	53
E. foetida (Manure Worm)	11	11	55	66
TOTAL	117	27	92	119
AVERAGE OF TWO SPECIES	58.5	13.5	46	59.5

(Adapted from Evans and Guild, 1948, and Wilcke, 1952)

development in from 100 to 125 days for two species which is less than that stated for ideal conditions. Michon[5] also reported that the manure worm (*E. foetida*) developed to full maturity in 66 days at 64.4 °F and in 45 days at 82.4 °F which, again, is less than that stated for ideal conditions.

Life Span

No one knows for sure the life span of an earthworm: in a favorable environment, an earthworm will live for many years. One observation was performed over a period of fifteen years, and the earthworms in the experiment appeared as young as ever. Other observations have calculated the life span in protective cultures as ranging from four to ten years. However, under natural field environments, it is doubtful if the life span of an earthworm exceeds two years.

6. Respiration

(See Fig. 2-10, Vol. 1.) Animals well adapted for land life have a heavy, impermeable skin which prevents excessive drying, but it also prevents respiratory exchange through the skin. In such animals, oxygen reaches the internal tissues by means of special respiratory devices, such as lungs.

While earthworms are terrestrial, or land, animals, they have not really solved the problems of land life; they have merely evaded them by restricting their activities to a burrowing life in damp soil, by emerging only at night when the evaporating power of the air is low, and by retreating deep underground during hot, dry weather.

Earthworms breathe by surface respiration in the same way as their aquatic ancestors; that is why they can live completely submerged in water, yet will die if dried out for a time. The outmost layers of the earthworm are thin and must be kept moist so that respiratory exchange can occur by diffusion through the general body surface, which is underlain by capillary networks. The small blood capillaries, which are close to the cuticle (see Fig. 2-12, Vol. 1), or body covering, allow the circulating blood to obtain oxygen and give off carbon dioxide through the moist body surface.

Moistening of the body surface is accomplished by mucous glands which occur in the epidermis and also by the coelomic fluid which issues from dorsal pores located in the middorsal line in the setae, or grooves, between segments.

A pigment which is generated in the coelomic fluid or waste material and accumulates in the body wall probably serves to shield the underlying tissues from the light, particularly the *ultraviolet,* which is very harmful to earthworms. One hour's exposure to strong sunlight causes complete paralysis in some earthworms, and several hours' exposure is fatal. Additionally an earthworm's body covering must remain moist for an earthworm to breathe through its respiratory system. If the body covering dries up, the earthworm cannot breathe.

This is thought to explain the death of many of the earthworms seen lying in shallow puddles after a rain. They have *not* been *drowned* by the water, as many people suppose, for earthworms can live many hours completely submerged in water. However, during a rain, the water that fills their burrows has filtered down through the soil and therefore contains very little oxygen. This forces some of the worms to come to the surface, where they are injured by the light and after a time can scarcely crawled. Additionally, if the earthworm's burrow is sealed off by water, carbon dioxide given off by the earthworm's respiratory system begins to build up. The carbon dioxide forms a weak acid with the water, forcing the earthworm out of its burrow. It is thought that the calciferous glands provide the earthworm with a degree of resistance to carbon dioxide for short periods of time. They probably remain in the rain puddles because of the protection afforded by the layer of water. Many of the dead worms seen after a rain were no doubt sick beforehand, perhaps as the result of heavy infestation with parasites; their death has only been hastened by the rain.

Aquatic Ability

With regard to the preceding statements on the earthworm's aquatic ability, 85 percent of an earthworm's fresh weight is water, with the coelomic fluid and blood comprising the largest part of the water weight. Earthworms in the soil are not fully hydrated; that is, they do not contain the maximum percentage of water. When earthworms are submerged in water for a period of time, their weight will increase by approximately 15 percent. When they are returned to the soil, they will dehydrate, or lose this additional water gain. An earthworm can lose up to 70 or 75 percent of its body weight without dying and up to 18 percent of its water weight without seriously affecting its ability to move or burrow.

Studies have shown that an earthworm, drawing on its ancestral aquatic capabilities, can survive as long as 31 to 50 weeks in aerated water by exchanging water and carbon dioxide via its surface respiratory system. Even then, it is suggested that the limiting factors are lack of feed rather than the drowning effects of water. It is also reported that earthworm capsules will hatch under water and that the young earthworms will feed and grow in the water. While earthworms have this aquatic capability, most species prefer moist soils to waterlogged or dry soils and will generally migrate from soils with either extreme condition.

Oxygen Consumption

Experiments have shown that earthworms have a daily rhythmic consumption of oxygen and that the rate of consumption is maximum at 6 a.m. and 7 p.m. These are not necessarily the periods of times that the earthworms are most active since it is possible that earthworms may be able to accumulate oxygen for future use. It is speculated that this accumulation of oxygen is used to form lactic acid, which is later converted to glyconol. This oxygen accumulation may also be the method by which earthworms can live for many hours without atmospheric oxygen. That is, under anaerobic respiration conditions, lactic acid is formed along with other compounds from the stored oxygen.

The respiration rate of an earthworm usually does not decrease until there is a very low oxygen pressure. Respiration decreases by 55 to 60 percent at an oxygen pressure of 38 millimeters of mercury, which is 25 percent of normal. This low level of oxygen normally causes a corresponding build-up in carbon dioxide. Generally, the earthworm's respiration is normal until the soil atmosphere contains approximately 50 percent carbon dioxide. However, in most cases, an earthworm will leave its burrow when carbon dioxide concentrations exceed 20 to 30 percent.

Earthworms use more oxygen per pound than does man. With a good supply of air, they are lively and colorful. As the oxygen is depleted, they become less active and their skins turn dark as blood surfaces to pick up oxygen that is not there.

Color Variations

The previously mentioned pigment accumulation in the body wall also explains difference in coloring of various earthworms of the same species when fed different feeds. It has been noted that the same species is deeper red, or maroon, when fed rabbit manure than when fed dairy or horse manure. Those fed a diet of horse manure are whiter than those fed dairy manure. It is not known if this is a result of a higher nitrogen content in the rabbit manure or of an excessive acid buildup in the beds caused by urine in unleached (i.e., unwashed thoroughly to remove urine) rabbit manure.

The whiteness of earthworms fed straight horse manure is probably due to several factors. Horse manure has a lower nitrogen content and is generally looser because of the greater amount of hay and straw. The looser horse manure provides a more porous area and subsequently allows more oxygen surface area within the area. The greater amount of oxygen across the surface respiratory area of the earthworm allows more oxygen into the respiratory system and more efficient conversion of nitrogen, phosphorus, and potassium elements in the feed to nourishing elements and oxides for elimination rather than pigment accumulation.

7. Digestion

Digestion is highly important to the functioning of the earthworm in the soil. It is the digestive system which "operates" on the feed intake of the earthworm, provides the earthworm with the necessary nutrients, and treats the food with enzymes which affect the casts (refer to Chapter 8) produced from the excretory system.

As the earthworm moves through the soil, large portions of soil and organic matter are swallowed; the organic debris and small organisms are separated from the organic soil particles, and are then digested. The earthworms obtain nutrition from the organic material in the form of plant material, animal remains and droppings, nematodes, protozoa, fungi, bacteria, and other soil micro-organisms. However, some species (*L. terrestris*) feed directly on leaves and even show a preference for certain types and conditions of leaves. Therefore, the digestive process will vary between species with respect to the enzymes and chemical compounds generated during the digestive process as a function of the organic material and minerals which are ingested. The following process is for the species *E. foetida,* or common manure worm.

The digestive tube of the earthworm is divided into different regions with specialized functions. In the earthworm, food enters the mouth and is swallowed by the action of the muscular pharynx, or throat. The pharynx is a muscular organ that mixes the food with a moistening secretion and pushes it farther down the tube through the narrow esophagus to the crop.

Pharynx or Throat

A ductless pharyngeal gland secretes an acid mucus into the pharynx containing an amylase, an enzyme which accelerates the hydrolysis of starch and glycogen to maltose (crystalline sugar, $C_{12}H_{22}O_{11}$). In *E. foetida,* this acid mucus probably does not contain any proteolytic enzymes which allow conversion of proteins to simpler and more soluble forms. However, there are reports that proteolytic enzymes are secreted by some species.

Calciferous Glands

The esophagus has three swellings on each side. These are the calciferous glands which excrete amorphous calcium carbonate (chalk) particles coated with mucus into the esophagus. The absolute function of this excretion is not known. Some researchers believe it controls the pH of the intestinal fluid, some that it regulates the calcium level of the blood, and some that it is responsible for excreting excess calcium carbonate assimilated from the soil. The more likely is that it is used to dispose of any excessive calcium obtained from the various salts present in the food. It is believed that these particular glands allow the earthworm to turn a small amount of highly acid organic matter into almost basic or neutral excretions (castings)—i.e., calcium carbonate tends to neutralize acids. While the earthworm has this capability, it is is limited and earthworms themselves can neutralize only a small amount of acids.

Many authorities associate the calciferous glands with the earthworm's resistance to carbon dioxide, because calcium readily combines with carbon dioxide. It seems likely that excess carbonic acid in the blood of the earthworm is combined with the calcium in these glands and is excreted as chalk by the intestine.

Crop

The esophagus leads into a large, thin-walled sac, the crop. The crop stores the food until the action of enzymes and bacteria, controlled by the weak calcium solution from the calciferous glands, breaks it up fine enough to enter the gizzard.

Gizzard

The gizzard is a sac with heavy muscular walls which secretes a form of pepsin to help digest and combine proteins and starches. The muscular walls of the gizzard, aided by mineral particles, sand, and very small stones swallowed by the worm grind the food into smaller pieces. These small stones, sand, etc., are the "teeth" of the earthworm.

Intestines

From the gizzard the food is transported by muscular action through the intestine, which continues almost uniformly to the anus, where the undigested food is excreted as castings. In the intestine, the food is digested by juices excreted by gland cells of the epithelium lining. In *E. foetida,* these juices consist of one amylase and two proteolytic enzymes. It is also thought that other enzymes may be created by symbiotic bacteria and protozoa.

The roof of the intestine dips downward as a ridge or fold. This is the typhlosole, a fold which increases the digestive surface that comes into contact with the intestinal contents by as much as five times.

The undigested food matter is stored in the intestine until it is excreted. Here it is thought that the undigested matter is enveloped by the peritrophic membrane which lines the intestine and covers the casts when it is excreted by the earthworm.

The intestine is the major organ of absorption, and small food particles which can be absorbed by the body are passed through the walls together with liquids. In the earthworm, which has a blood circulatory system, the intestinal walls are lined with many capillaries.

The proteins, sugars, and other cellular food molecules are absorbed into the blood capillaries of the intestinal walls and are distributed to the body cells. The waste matter brought from the tissues is passed through cilia in the excretory system, a crude form of kidney (two of which are located in each segment), to the outside of the skin in the form of mucous, which acts as a lubricant for the earthworm as it burrows its way through the soil.

Although the earthworm has a fairly complicated digestive apparatus, it does not include teeth nor does it include acid as a strong glandular secretion, as in mammals, that can be used to convert proteins and carbohydrates into energy, or food value. The earthworm can, and will, ingest particles larger than is generally thought, but these are swallowed to the crop and gizzard, where they are acted on only by a small amount of enzymes.

Enzymes

The composition of enzymes is still somewhat of a mystery to scientists, but it is known that they act on organic and cellular matter to produce digestive substances such as lipase, protease, and rennin, among other things. *Lipase* is a digestive substance also found in cows and other ruminants. *Protease* is a form of incomplete protein found in vegetables and grain. *Rennin* is a form of vegetable "milk," or a semifluid mass of converted starches and proteins. All of these chemical actions and reactions are necessary to transform inorganic forms into organic (carbon) forms, of which all life forms are composed and, which are necessary for energy. Each living organism is a group of millions of tiny cells, each of which must be continually added to or replaced so that metabolism may continue. When food is lacking, metabolism and growth are reduced.

As stated, the earthworm's digestive tract generates few of the enzymes necessary to create the bacterial action required to convert proteins and carbohydrates. Converted, or reconstructed, proteins are essential to all animal metabolisms and the earthworm cannot live without them. Earthworms can and will starve to death (die of malnutrition) in the presence of feed if bacteria, or ordinary germs, do not process it for them.

It is thought that the earthworm's digestive tract generates enzymes (such as cellulase) that allow it to convert cellulose and carbohydrates (sugars and starches) into fattening food values and energy faster than it can a protein. Since most proteins must be broken down by a bacterial process before they can become ready earthworm food, the ideal earthworm feed should have a higher cellulose and carbohydrates content than protein content.

All of this means that the earthworm, which has a digestive system which excretes few digestive fluids, must depend on external bacterial or decay germs for predigestion of its food. This process is commonly known as fermentation or souring. Bacteria manufacture, or generate, small amounts of acid to dissolve their own food.

8. Coelom

The coelom, or body cavity is a fluid-filled space between the digestive tube and the body wall of the earthworm. The coelom is completely lined by a sheet of mesoderm cells, the coelomic lining, or peritoneum. The coelom, which separates the intestines from the body-wall muscles, is filled with a fluid which contains ameoboid (amoeba-like) cells and many other dissolved substances. The coelomic fluid bathes all of the internal organs and thus serves a role similar to that of the circulatory system, even though it has no direct connection with that system. The coelom also plays a role in excretion and reproduction.

Sheets and bands of mesenteric membranes, located between the gut and body wall, form pouches and divide some of the internal organs into separate areas. In the area around the intestine, some peritoneal cells are changed to form chloragogenous tissue. These cells contain chloragosomes, which are yellow or greenish-yellow globules.

The coelomic fluid is normally a milky white substance;

however, it may sometimes contain cells of oil droplets, called *eleocytes,* which may make the coleomic fluid yellow in color. The coelomic fluid will differ in consistency between earthworm species and with the soil moisture. The fluid will be thinner and less gelatinous for earthworms in a moist environment than for earthworms in a dry environment.

Many different types of particles are suspended in the coelomic fluid. Most of the inorganic materials in the fluid are calcium carbonate crystals. However, the fluid also contains phagocytic (waste eating) amoebocytes which feed on waste materials, vacuolar lymphocytes, and mucocytes. The latter provides the fluid with a mucilaginous component. Other products in the coelomic fluid include corpuscular bodies, bacteria, and parasites such as protozoa and nematodes. The disintegrated remains of setae, amoebocytes, and nematode cysts may be found in the posterior end of the earthworm. These are modules of aggregated dark-colored masses, known as "brown bodies."

9. Excretion

The earthworm has the same problem of excretion as other living things. Water balance must be maintained and nitrogen-containing waste products must be eliminated. The excretory system is segmentally arranged. A pair of excretory organs, called *nephridia,* are located in each segment except the first three, and the last. Similar organs are known as kidneys in man and are the main organs of nitrogenous excretion.

Each nephridium really occupies two segments. Each organ consists essentially of a tube which opens at one end by a pore on the ventral surface and internally on the other end by a ciliated funnel. This funnel lies in the coelom of the segment anterior to the segment containing the nephridium and its external pore. Wastes are extracted from the blood which passes through the nephridium and also from the coelomic fluid with which the organ communicates via its ciliated funnel. Microscopic particles in the coelomic fluid are wafted into the funnel opening by beating of the cilia around the edges of the funnel. These wastes are swept to the excretory pore opening by means of cilia lining the nephridium and by waves of muscle contraction in the wall of that portion of the nephridium which leads to the excretory pore. The waste material from the nephridia, via the excretory pores, is largely responsible for the "slime", or coating of moisture (mucous) that is always found on the skin of a healthy earthworm.

The "slime," or mucous, acts as a lubricant as the earthworm moves through the soil and forms a protective coat against toxic materials entering the body via respiratory exchange. The mucous, which contains a large amount of nitrogenous material, also binds together the soil particles, which form the earthworm's burrow. It is estimated that this mucous constitutes approximately one-half of the nitrogen which is excreted each day by the earthworm.

The nephridia are not the only means of excretion in the earthworm. The coelomic lining surrounding the intestine and the main blood vessels is modified into special chloragen cells. Wastes extracted from the blood accumulate in the chloragen cells, which finally become detached and float in the coelomic fluid. Some of the chloragen detritus (debris) is removed by the nephridia as previously noted. Some of it is engulfed by the ameboid cells of the fluid which finally wander into the tissues and disintegrate, leaving the wastes as a deposit of pigment in the body wall.

Other cells within the body of the earthworm may discharge *urea* and *ammonia* into the coelomic fluid for later elimination. In addition to being eliminated by the nephrida, the chloragogen cells in the coelomic fluid can be assimilated by amoebocytes. These may then be either deposited in the coelomic fluid in bulky nodules, or deposited in the body wall. Some amoebocytes may also be in the blood. These are generally deposited in the intestinal wall and then fall off into the intestine where they are excreted with the castings. Additonal excretory cells, possibly small uric acid crystals, function similarly to the amoebocytes.

Edwards and Lofty[6] state, "The nephridia obviously acts as a differential filter, because there is much more urea and ammonia, but less creatinine and protein in the urine they produce, than in the coelomic fluid. The composition of the urine changes as the fluid passes along the nephridium (Ramsay, 1949), but it is not known whether osmotic changes (changes in osmosis, author) are due to resorption of salt or secretion of water. Certainly, the urine is at a much lower osmotic pressure than the coelomic fluid, but how this is achieved is not clear. There is some evidence that granules of waste material can be taken up by the ciliated tube of the middle wall and remain there, so that these parts of the nephridia may act as kidneys of accumulation. Bahl (1947) concluded that nephridia have three functions in excretion, namely filtration, resorption, and chemical transformation. He believed protein was reabsorbed through the nephridal wall against a concentration gradient, but this remains to be confirmed and was not accepted by Martin (1957)."

A diagram presented by Edwards and Lofty to summarize one possible mode of nephridia functioning in the excretion and resorption process is shown in Figure 3.6.

Luminescence

Some earthworms produce a short-lived phosphorescent or luminescent substance which appears as spots or blotches on their bodies or in the "slime" trail behind the earthworm. This luminescence, which generally fades in approximately 30 seconds, is probably a slime excreted from the prostomium or it may be caused by coelomocytes in the slime, or mucous, which the earthworm excreted through its body pores. The exact function of this luminescence is not known but it is thought that it is a protective mechanism since it occurs in response to irritation, vibration, etc.

10. pH And Acid Sensitivity

Vol. 1 stressed the fact that earthworms, because of their limited digestive system, require a certain amount of acidity for "predigestion" of their food, but that too much acidity could be detrimental to the earthworms or drive them from the beds. The same factors hold true for earthworms in the soil, except that some species of earthworms in the soil can stand a more acid soil than the commercially raised earthworms under discussion in Volume 1. However, almost all earthworms prefer a soil with a near-neutral pH and, to encourage earthworms in the soil, the soil should be maintained between pH 6.0 and pH 7.2. While this is somewhat more acid than that recommended for the bedding used to raise commercial earthworms, it encompasses the range over which most plants will grow, as shown in Figure 13.3 and studies have shown that earthworms will thrive in this soil range.

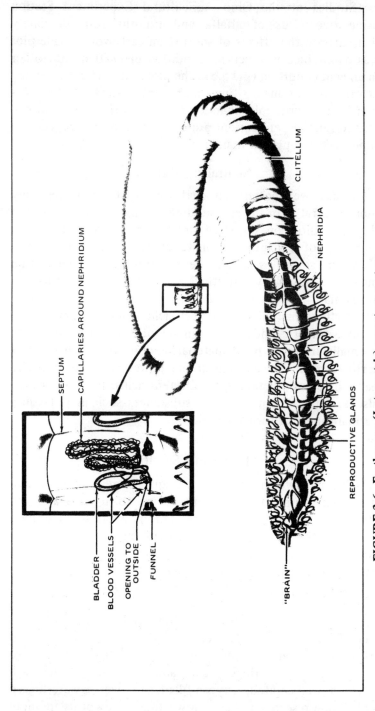

FIGURE 3.6—Earthworm (L. terrestris) excretory system

Studies at the Ohio Agricultural Experiment Station using two plots of alfalfa and timothy sod adequately demonstrate the effects of soil pH on earthworms. One plot was maintained in a neutral condition pH 7.0) and the other in an acid condition (pH 5.5). The plot which was maintained at pH 7.0 contained 3 times more earthworms and 2.6 times more earthworm holes than the plot with pH 5.5. Additionally, the relative infiltration rate of water into the soil was 7.5 times faster at pH 7.0 than at pH 5.5.

Neutralizing Effect

It has been noted that earthworm castings are generally more neutral than the surrounding soil and, as pointed out in the discussion on digestion, may be due, in part, to the secretion of calcium carbonate by the calciferous glands. However, some researchers believe this neutralizing action may be caused by intestinal secretion and excretion of ammonia.

Earthworms can also detect acids such as citric, phosphoric, tortoric and malic, among others. These acids are normally found in plant materials and are tolerated by most earthworms in low concentrations; however, different species have different tolerances to different acids. Earthworms are also sensitive to many other substances such as potassium permanganate or formaldehyde.

11. Moisture Sensitivity

The earthworm, originally an aquatic creature, requires a moist environment to prevent injury to its delicate skin structure and for proper body functions. Section 6 describes, in detail, the oxygen requirements of earthworms and their aquatic ability. If a soil is too dry, earthworms will either burrow deeper into the soil, go into hibernation with a subsequent loss in body weight, or die. Under long periods of drought, most of the earthworm population in the area will die and, under favorable conditions, it has been estimated that it can take up to two years for the area to repopulate to its original density. Conversely, earthworms will tend to migrate from waterlogged soils and, if given a choice, will select a well-aerated, oxygen-filled soil over a waterlogged soil. Some researchers have shown that earthworms are more numerous and active and produce more capsules in soils with a moisture content between 15 and 30 percent.

12. Temperature Sensitivity

Commercially raised earthworms are most active when bed temperatures are maintained between 60° and 70°F. They remain highly active in bed temperatures up to 80°F; however, reproduction and hatching out of egg capsules decrease above 70°F. An earthworm will die if exposed to freezing or lower temperatures, but may live and continue to breed with bed temperatures up to 100°F or higher in well-shaded locations if plenty of moisture is always available.

The preceding statements were made in Volume 1 in regard to commercially raised earthworms. Studies[8] have shown that these statements are also true for earthworms in the soil. The activity of earthworms, their metabolism, respiration, reproduction, and growth are all affected by temperature.

In most cases, they will burrow down below the frost or heat line. Even though the surface soil temperature may be above the extreme in which earthworms can live, they can maintain their body temperatures below that of the surrounding soil by evaporation of water from their body surfaces. Earthworms can also acclimate themselves to gradual temperature change above or below their normal range. This explains their ability to survive seasonal changes.

It is hoped that this detailed chapter on the body of the earthworm will assist you in keeping your earthworms healthy and multiplying.

REFERENCES

1. Edwards, C. A. and Lofty, J. R. (1972) **BIOLOGY OF THE EARTHWORM**, Chapman & Hall Ltd., London, England and available from Bookworm Publishing Co., p. 138.
2. Ibid. pp. 58-61.
3. Ibid. pp. 58-60.
4. Ibid. p. 61.
5. Ibid. p. 130.
6. Ibid. pp. 83-84.
7. Ibid., p. 84.
8. Ibid. pp. 129-133.

CHAPTER FOUR

EARTHWORMS IN THE SOIL

A. THE SOIL AFFECTS EARTHWORMS

Various species of earthworms are widely distributed in all areas of the world. Generally, greater populations and varieties of species are found in humid areas than in dry or arid areas. However, earthworms will and do populate the waterways and irrigated areas of desert regions. This fact is illustrated by the large number of earthworms found along the Nile River area of Egypt, an area which has had fertile soil for thousands of years of agriculture. There are many different factors which will affect the earthworm's ability to populate a soil: soil type, food supply, organic matter content and environmental factors such as moisture, temperature, pH, aeration, and carbon dioxide build-up. The soil type, food supply and environmental factors can result in wide variations in the size and species of the earthworm population. Populations can vary regionally because of climate and locally because of the soil type and the quality and the quantity of organic matter in the soil. Population differences may occur in adjoining lands because of agricultural practices, such as crops planted and the use of artificial fertilizers and pesticides.

Only some of these factors affecting earthworms in the soils have been studied to any great extent by the scientific community. Not all of the findings are in agreement. Since many of the different earthworm species have different tolerances for different environments, it is speculated that many of those findings which are not in agreement may be due to different species being studied. Each of the factors affecting earthworms in the soil is discussed in the following sections.

1. Soil Type

Unfortunately, this is one of the factors which have received very little attention by the scientific community and the experts tend to disagree slightly on the type of soil that is most beneficial in attracting and maintaining an active earthworm population. But they know that some are better than others.

Charles Darwin discussed the lack of study and proper classification of earthworms in Victorian times. He noted that there were only 8 acknowledged species under study in Scandinavia and Germany, but the earthworms did survive in such different soils as fresh water swamps (not salt water) to rocky highlands.

Dr. McIntosh, however, found worm-castings at a height of 1,500 feet on Schiehallion in Scotland. They are numerous on some hills near Turin at from 2,000 to 3,000 feet above the sea, and at a great altitude on the Nilgiri Mountains in South India and on the Himalaya.

Earthworms must be considered as terrestrial animals, though they are still in one sense semi-aquatic, like the other members of the great class of annelids to which they belong. M. Perrier found that their exposure to the dry air of a room for only a single night was fatal to them. On the other hand he kept several large worms alive for nearly four months, completely submerged in water. During the summer when the ground is dry, they penetrate to a considerable depth and cease to work, as they do during the winter when the ground is frozen.[1]

Dr. Henry Hopp[2] of the U.S. Department of Agriculture states,

> Where reasonable moisture conditions prevail, their (earthworms) occurence is determined by soil variation. They are more common in soils derived from limestone or otherwise rich in plant nutrients, than in shale or outwash soils. Soil texture influences the earthworm population. *Sandy soil contains fewer earthworms than clay soils.* This is fortunate because sandy soil is likely to have good structure naturally, while clay soil packs together and becomes too hard for crop growth unless agencies like earthworms are present to keep the soil granulated.

This statement is not in complete agreement with statistical data presented by Edwards and Lofty[3] who state,

> Guild (1948) made a survey of the main soil types in Scotland, and reported that there were differences both in total numbers and relative numbers of each species. Light and medium loams had higher total populations of worms that heavier clays or more open gravelly sands and alluvial soils. *A. caliginosa* was the dominant species in all soil types, but *A. longa* was less important in open soils, gravelly sands and alluvial soils. In a survey of the distribution of earthworm

species in the Hebrides, Boyd (1957) compared the relative abundance of earthworm species in light soils with those in calcareous sand and dark peaty soils. Six species were more abundant on the light soils and six on the dark ones. In particular, *A. caliginosa* and *L. castaneus* were much more numerous in the light soils, and *B. eiseni* and *D. octaedra* in the dark soils.

A few small species of earthworms can survive in deserts and semi-deserts (Kubiena, 1953; Kollmannsperger, 1956), and some worms can inhabit the arid, cold soil of north-east Russia. It seems that although earthworms do better in good soils rather than poor ones, they can survive in many different kinds of soils providing there is adequate food and moisture.

The statistical data presented by Edwards and Lofty (see Fig. 4.1 and Table 4.2) shows that light and medium loamy soils attract a greater earthworm population than a clay or acid soil. Additionally, Figure 4.2 shows that two species, *A. caliginosa* and *A. longa* prefer a clay soil over a sandy soil, two other species, *L. rubellus* (red worm) and *L. terrestris* (nightcrawler) favor a sandy soil over a clay soil.

The preceding data seems to indicate that different species can survive in different soil types, provided adequate food and moisture are present, but all species prefer a light-to-medium loamy soil which is normally loose, well aerated, holds moisture, and has a great deal of organic matter. These same factors were stressed in Volume 1 as those necessary to successfully raise earthworms on a commercial basis and it appears that the same criteria apply to earthworms in the soil.

The Staff of **Organic Gardening and Farming Magazine** on page 483 stated,

> Earthworms can work in any kind of soil, even clay, but in that case, it will take longer for the worms to cultivate it thoroughly. Ordinary earthworms will thrive only in the kind of soil they are born in. In taking some of these creatures from a sandy soil and transplanting them to a very heavy clay loam, most of them will die off. *Where earthworms are bred in boxes from a strain of worm that has long lived in captivity, it will be found that they will live in almost any kind of soil.*

The previous statement is of extreme importance to those who raise earthworms on a commercial basis since it was made by persons who do not raise earthworms on a commercial basis. It is believed that Dr. Thomas J. Barrett

TABLE 4.1 – Relations of soil type to earthworm population

Soil type	Population		No. of species
	(Thousands/acre)	(Number/m^2)	
Light sandy	232.2	57	10
Gravelly loam	146.8	36	9
Light loam	256.8	63	8
Medium loam	226.1	56	9
Clay	163.8	40	9
Alluvium	179.8	44	9
Peaty acid soil	56.6	14	6
Shallow acid peat	24.6	6	5

FIGURE 4.2 – Density of earthworm population in various soil types in Scotland

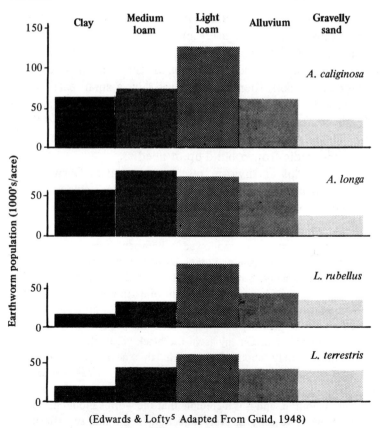

(Edwards & Lofty[5] Adapted From Guild, 1948)

originally pioneered the raising of manure and red worms in captivity for use in gardening and farming. His work, and that of others, have indicated that the commercially raised manure and red worms can be transplanted to almost any kind of soil and that they will thrive and follow the yearly cycle shown in Figure 3-1. The success of transplantation is, of course, dependent on environmental factors such as food availability, moisture, aeration, and temperature, etc. These claims have been disputed by many people, both scientific and nonscientific. However, chapters of this book will substantiate the claims that the commercially produced earthworms can be transplanted, and are beneficial to the soils to which they are transplanted.

2. Food Supply

One of the major factors which affect the distribution and population of earthworms in the soil is the availability of organic matter and mineral soil as a food source. The type and amount of organic matter and minerals present in the soil influences the earthworm population, the species, the growth rate, and the reproduction rate. If there is little organic matter in the soil, there will generally be a small earthworm population. Also, if the earthworm population is small, the organic matter will decay on top of the soil and form a thick mat. This happens in cool areas of evergreen forests where the soil is acidic from a build-up of needles.

The famous earthworm researcher, Charles Darwin, was the first to note that earthworm castings were evidence that the earthworms were ingesting soil each night, not just making burrows:

> If the earth were swallowed only when worms deepened their burrows or made new ones, castings would be thown up only occasionally; but in many places fresh castings may be seen every morning, and the amount of earth ejected from the same burrow on successive days is large. Yet worms do not burrow to a great depth, except when the weather is very dry or intensely cold. On my lawn the black vegetable mould or humus is only about 5 inches in thickness, and overlies light-coloured or reddish clayey soil: now when castings are thrown up in greatest profusion, only a small proportion are light-coloured, and it is incredible that the worms should make fresh burrows daily in every direction in the thin superficial

layer of dark-coloured mold, unless they obtained nutrient of some kind from it. I have observed a strictly analogous case in a field near my house where bright red clay lay close beneath the surface.

Two great piles of leaves had been left to decay in my grounds, and months after their removal, the bare surface, several yards in diameter, was so thickly covered during several months with castings that they formed an almost continuous layer; and the large number of the worms which lived here must have subsisted during these months on nutritious matter contained in the black earth.

The lowest layer from another pile of decayed leaves mixed with some earth was examined under a high power [magnifying lens], and the number of spores of various shapes and sizes that it contained was astonishingly great; and these crushed in the gizzards of worms may largely aid in supporting them. Whenever castings are thrown up in the greatest number, few or no leaves are drawn into the burrows; for instance, the turf along a hedgerow, about 200 yards in length, was daily observed in autumn during several weeks, and every morning many fresh castings were seen; but not a single leaf was drawn into these burrows. These castings from their blackness and from the nature of the subsoil could not have been brought up from a greater depth than 6 to 8 inches. On what could these worms have subsisted during this whole time, if not on matter contained in the black earth? On the other hand, whenever a large number of leaves are drawn into the burrows, the worms seem to subsist chiefly on them, for few earth-castings are then ejected on the surface. This difference in the behaviour of worms at different times, perhaps explains a statement by Claparede, namely, the titurated leaves and earth are always found in distinct parts of the intestines.

Worms sometimes abound in places where they can rarely or never obtain dead or living leaves; for instance, beneath the pavement of well-swept courtyards, into which leaves are only occasionally blown. My son Horace examined a house, one corner of which had subsided; and he found here in the cellar, which was extremely damp, many small wormcastings thrown up between the stones with which the cellar was paved; and in this case it is improbable that the worms could ever have obtained leaves.[7]

Darwin also noted that worms would eagerly eat raw meat, fat and dead worms (all protein sources) as well as the cellulose sources of leaves and decaying roots.

a. Studies

Earthworms thrive on a nitrogen-rich diet, thus they prefer animal decaying matter to plant material. Earthworms do well in various manures. One test on two fields that had been permanently cropped to wheat was recorded as follows:

> One of these plots received approximately 15 tons of manure per acre while the other was unmanured. The earthworm population was three to four times greater in the manured plot than in the unmanured plot. In other observations, the earthworm population was up to 15 times greater in annually manured plots than in unmanured plots.
>
> Other experiments and observations have shown body-weight gains of up to 18 percent when earthworms were fed clover leaves for 40 days, 71 percent when fed manure on the surface, and 111 percent when fed manure incorporated into the soil. Weight losses were reported when earthworms in the same experiment were fed various other roots and leaves.[8]

Similar, but less dramatic, changes in body weight were observed in experiments by the United States Department of Agriculture, when the common field worm was fed different kinds of organic material combined with a silt-loamy topsoil. In this experiment, the greatest gain, 21.5 percent, occurred when the earthworms were fed lespedeza leaf litter and 10.7 percent when fed fresh cow manure. These tests substantiate the fact that earthworms will thrive better on organic matter high in nitrogen and cellulose.

Additionally, large earthworm populations may be found in fields with a high percentage of dead roots and other organic matter. A corresponding decrease in earthworm population will occur when the land is plowed and planted in crops.

Earthworms can tell one leaf from another. They tend to avoid those leaves with resins like pine needles or ones with strong aromatic content. Certain leaves need to be leached by months of rain before earthworms will eat them. Darwin also noted that earthworms will nearly always grasp a leaf by its pointed end or smaller end, thus dragging it the most efficient way into its burrow for later meals.

In Chapter 4 of Volume I, there is a detailed discussion of proper feeding of domesticated earthworms. They need a balanced diet just as we do. Now the ideal feeds, that vermiculturalists use in raising their fat worms, will give the gardener some ideas of how best to keep his worms fed.

Selecting a feed and feeding earthworm beds is really no different that feeding any other livestock. Some feeds are naturally better than others because they contain more vitamins, minerals, proteins and other elements which can be easily converted by the earthworm into the nourishing food values essential for health and rapid growth. Feeding methods will vary somewhat with the type of food.[9] This Chapter 4 discusses the values and hazards of these feeds:

a.) animal manures
b.) cardboard
c.) nut meals
d.) waste products and compost
e.) commercial feeds
f.) wood and paper products
g.) sewer sludge

Earthworm feeds should also contain the water-soluble vitamins (B, C, etc.) as well as minerals such as calcium, phosphorus, potassium, sodium, etc. These vitamins and minerals are as essential to the health, well being, and body functions of the earthworm as they are to any animal or, indeed, human being.

The ideal earthworm feed, as previously stated, should have a higher cellulose and carbohydrate (starch) content than protein content. Cellulose is the chief component of the solid framework (cell walls) of plants. Cellulose is a carbohydrate ($C_6H_{10}O_5$) which is convertible into glucose (sugar) by hydrolysis (addition of water). Almost all leaves, vegetable products, shells, husks, and chaff (bran) are high in cellulose. Grain contains protein and starches. All of the cereal grains—corn, oats, wheat, barley, millet, etc.—are high in starches and sugars. Oats, wheat, and corn also have a high protein content. The shells, husks, and chaff (bran) of all grains have a high cellulose and natural vitamin (A, B, C) content. The cellulose in all kinds of bran or chaff becomes available as tissue-building food as soon as oxidation and bacteria break it down, or immediately if it is ground into a fine flour which can be easily assimilated by the earthworms.

Legume hays, including alfalfa, clover, lespedeza, cowpea, vetch, and peanut, are high in protein (12 to 20 percent) as well as cellulose, vitamins, and minerals. Grass hays, including timothy, prairie, Johnson grass, sudan grass, and carpetgrass contain about half as much protein but are rich in cellulose, vitamins, and minerals. The problem with most hays such as these is that they decay too slowly to provide immediate benefit. However, if they are composted and used as bedding and feed, or if they are chopped or ground and mixed with the heavier feeds (manures, etc.), they will keep the bedding or feed loose and provide a constant supply of feed as they decay.[10]

Use your common sense in what you spread on your garden, remembering that earthworms like a fairly neutral pH, moderate temperatures and moist ground. But avoid over watering, as earthworms become sluggish in very wet conditions. Also adding some limestone will help keep the acidity of your soil under control.

b. Amount of Food Consumed

Contrary to some observations, earthworms do consume a large amount of food. Edwards and Lofty[8] report that a study by Guild (1948) showed that mature individuals of *L. rubellus,* the common red worm, could ingest 16 to 20 grams (0.56 to 0.7 ounces) *dry weight* of manure per year. If there are approximately 1200 mature worms per pound in the soil as it has been estimated then each individual weighs approximately 0.0133 ounce (1/100 oz.). (It should be noted for clarification that a mature red worm is one that has developed to breeding age and its size in the soil will roughly equate to what is known as "bedrun" in the commercial industry.) This means that an earthworm can eat 11 to 14 percent of its body weight in *dry weight* food each day whereas a human normally eats approximately 4 percent of his body weight in *dry weight* food per day.

Note that the manure is measured in dry weight and *not* wet weight. The absorption weight of most manures (with dry straw, etc., mixed in) will run 7 or 8 times dry weight amounts. Earthworms will, then, consume 77 to 112% of their body weight in wet weight food per day.

Studies by the United States Department of Agriculture in 15 different geographical locations have found earthworm populations ranging from 4 to 50 per square foot, or 190,000 to 2,200,000 per acre in the first seven inches of soil. The average population for the 15 sites studied was 23 earthworms per square foot, or 1,004,000 per acre. Using the preceding ingestion calculations for *L. rubellus,* the average population for the 15 sites could consume 12.9 to 16.1 ounces per square foot, or 17.57 to 21.96 tons per acre, of *dry weight* manure per year. The red worm, being one of the smaller earthworms, naturally consumes less than other earthworms. Some estimates for other earthworms are up to 1.4 ounces *dry weight* per year, or more than twice as much as the red worm. If this were equated to 100,000 bedrun earthworms in a commercial 3 foot by 8 foot bed, each bed of earthworms could consume 1.8 to 2.2 tons of *dry* weight manure per year.

These scientific studies show how best to feed your earthworms to have active and large worms. With proper feed, your earthworms will thrive and improve your agricultural soils, both chemically and structurally. It is a worthwhile partnership.

3. Environmental Factors

The several environmental factors that affect the earthworm's ability to populate a soil are discussed in detail in other chapters in this book. Briefly, they are

    a.) a moderate soil temperature,
    b.) a reasonable soil moisture level,
    c.) a proper supply of oxygen and carbon-dioxide in the soil "atmosphere" or aeration, and
    d.) a near neutral pH level.

Now with these ideal agricultural factors arranged, the agriculturalist needs only to ascertain that his soil has a good supply of organic materials and of any minerals that my be lacking in that soil. Also the soil structure can be improved as needed, but mainly try to avoid those conditions that poison earthworms, such as a naturally too salty or too acidic soil condition *or* a man-made chemicallly hostile condition through the wrong use of pesticides or insecticides. Although we cannot control the seasons or much of our weather, we can improve many factors in our soil to make it closer to the ideal soil for our earthworm populations.

B. THE EARTHWORM AFFECTS THE SOIL

1. Effects on Soil Particle Breakdown

It has been observed in fields and pot experiments that the size of sand or other chemical particles in earthworm casts is smaller than those in the surrounding soil. However, the number of particles broken down was small and it is doubtful if it is significant compared to breakdown of minerals by weathering and chemical processes.

While the amount of minerals broken down by the earthworm may be insignificant in the overall process of soil formation, it is significant to the study of the earthworm; it is one more tool with which the lowly earthworm assists nature in her overall soil building and plant growth processes. To the best of the writers' knowledge, no one has speculated on how this breakdown occurs. It is probable that it occurs in the earthworm's gizzard which uses ingested mineral particles, sand, and very small stones to grind the food particles into smaller pieces. This grinding process, coupled with the weak acids and enzymes in the gizzards probably breaks the small "grinding" stones down into even smaller pieces.

2. Effects on Soil Particle Aggregation

Aggregates are small soil particles bound together into water-stable granules in such a manner that the soil will not crust or compact readily. Thus, aggregation allows rain and irrigation water to enter the soil easier, increases aeration, increases microbial activity, decreases erosion, and allows easier cultivation of the soil. Therefore, a soil which contains a large number of aggregates is more fertile than one with less aggregates, provided the required nutrients and minerals are present.

Water stability is one of the more important features of the soil. The binding material, or cement, which holds the soil particles together is largely organic in nature, and is produced during the biological decomposition of organic materials in the soil and, when dry, will not readily dissolve in water. This aggregation, or granulation, is different from that produced by cultivation as the granules produced by cultivation will immediately dissolve in water but those produced by aggregation will not. Cultivation is, therefore, only a temporary means of loosening and aerating the soil, while aggregation has a more long-term effect.

It is generally agreed that earthworm casts contain more water-stable aggregates than the soil in which they are found and that earthworms are one of the most effective means of increasing the water stability of soils. Experiments by Hopp (1946) showed that the percentage of aggregates in soils with earthworms was greater than that in soils without earthworms. In two soil plots, one with earthworms and one without, the soil with earthworms contained 6.1 percent more aggregates three days after the earthworms were introduced than the soil without earthworms. In another experiment, 13 different soils were incubated, one sample each with earthworms and one without. After one week, the 13 soils with earthworms tested an average of 9.65 percent more aggregates than the soils without earthworms. The exact duration of time over which earthworm casts remain water stable is not known but it is thought that these effects last for several weeks or a few months.

Exactly how the earthworm accomplishes this equal efficiency and the amount of aggregation and the stability of casts are dependent on the earthworm's behavior and the amount and type of organic matter in the soil. The latter is demonstrated by the fact that earthworm casts in forests or in lands planted to grass are greater in number than those produced by earthworms in lands planted to arable crops.

It is known, however, that aggregation occurs during decomposition of organic materials by microbial activity, and that the water stability of soils can be increased by calcium humate and polysaccharide gums, products of this decomposition. Additionally, studies have shown that there are normally fewer fungi in fresh earthworm castings, but that the fungi hyphae grow intensely for 15 days from the time the casts are excreted and then start to decrease. This intensive growth of fungi hyphae seems to correspond to changes which occur in earthworm castings; that is, earthworm castings tend to increase in water stability for 16 days and then start to decrease. Some of the nonconclusive theories put forth by researchers in the field are:

a. Aggregates are cemented together by secretions in the earthworm's intestines. However, if this were true, the casts produced by earthworms in forests or grasslands and in cropped land would be similar in water stability.
b. The earthworm's digestive system synthesizes calcium humate from the decomposing organic matter which is ingested. This substance and the calcium secreted from the calciferous glands cement the soil particles together. This is feasible since the amount of calcium

humate is controlled by the quantity and quality of organic matter in the soil. This could also account for the difference in the casts produced by earthworms in forests or grasslands and in cropped lands.

c. Bacteria in the casts produce water-stabilizing gums through microbial action on organic matter ingested by the earthworm. This also is feasible, using the same reasoning as in the preceding theory.

In all probability, the water stability of earthworm casts is caused by a combination of all of these effects. Calcium humate and polysaccharide gums synthesized by microbial activity in the earthworm's digestive tract provide a partial binding effect and then the fungi provide an even greater binding effect when the casts are excreted. It is also possible that a material similar to the slime or mucous, previously discussed, used to bind the soil particles in the earthworm's burrow may be used as a water-stabilizing element in the casts.

3. Effects on Soil Turnover

Earthworms, as they burrow and feed, swallow great quantities of organic matter and soil. They digest it, extract its food value, and excrete the residue. The earthworm takes organic matter from the soil surface and deposits it as castings in the lower soil levels and then takes soil from the lower levels and deposits it in the upper levels. In this manner, the earthworm continually turns the soil over, providing decaying organic matter rich in nutrients and minerals at a level where they can be used by plant roots, and builds a layer of fine stone-free topsoil rich in minerals from the rocks in the subsoil.

The amount of soil which is turned over varies greatly, depending on the soil type, environmental factors, and earthworm population. In some areas, it is estimated using Darwin's ratios that earthworms turn over 100 tons of soil, or more, per acre per year and build a fine topsoil layer at the rate of two inches per year. United States Department of Agriculture experiments have shown that in some areas earthworms convert 700 pounds of soil into castings per acre per day. These castings are generally mixed with the upper few inches of the native soil, causing a build up in topsoil.

Soil turnover by earthworms also has a mixing effect. In soils without earthworms, there are generally distinct horizontal layers of material; fresh organic matter, decaying organic matter, humus, and layers of progressively coarser soil materials (topsoil, subsoil, etc.). In soils with earthworms, the turn-over action generally mixes the layers so that they are indistinguishable. This action keeps the soil loose and more porous, providing better aeration, drainage and moisture retention in addition to distributing the organic matter and minerals throughout the soil for better utilization by plants.

4. Effects on Soil Moisture, Porosity, and Drainage

The preceding paragraphs have shown that earthworms have a definite effect on the aggregation of organic matter and soil particles, and distribution organic matter throughout the soil where it may be better utilized by plants. While these factors are important to the soil, the burrowing activities of the earthworms are of equal importance.

As explained in Chapter 12, most available soils consist of thin clay or silt particles which fit very tightly together, creating a soil which compacts very easily and is practically impervious to water. This type of soil is very slow to absorb water, causing runoff and soil erosion. Water which is absorbed is bound tightly in the soil particles, resulting in a soil which has poor drainage and aeration, and becomes waterlogged after a heavy rain or watering. In many cases, the compaction is so great that only the upper part of the soil is wet. When dry, this type of soil is almost impermeable to water or air and will harden to a thick crust.

Disregarding the other beneficial physical and chemical effects of the earthworms on the soil, their burrowing activities alone can improve the physical structure of the soil. As the earthworms burrow through the soil, they create interconnecting channeling extending several inches down into the soil and, in some cases, even into the subsoil. This allows water to penetrate throughout the soil rapidly, wetting the whole topsoil area by capillary action, and thus preventing runoff and erosion. Additionally, the aggregates formed by the earthworm casts hold the moisture for future use. The earthworm burrows are coated with a slime, or mucous, which keeps the holes open even when wet. In one experiment (Chapter 2), drainage was four to ten times faster in soils where earthworms were present than in soils where there were no earthworms. In another experiment, the initial rainfall (water) absorption rate of a soil increased from 0.2 inches per minute to 0.9 inches per minute after the soil had been worked by earthworms for one month. In all cases, soils with and without earthworms had approximately the same total water-holding capacity but most of the water ran off the soils without earthworms because of the slow infiltration rate.

Earthworms alone cannot increase the water absorption and drainage of a soil. If there is no organic matter in the soil, the earthworms will die. When there is organic matter in the soil, there is generally plant growth and the combination of earthworms and plant roots has a greater effect on water infiltration. Mulch, as noted in Chapter 5, can be highly

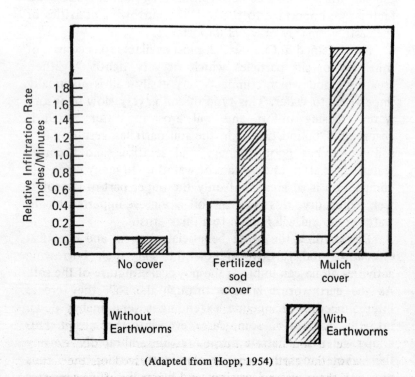

(Adapted from Hopp, 1954)

FIGURE 4.3—Effects of earthworms and cover on infiltration rates of clay soil

important in increasing the filtration rate of a soil in that it acts as a cushion for the water, allowing it to gradually soak into the soil. Organic mulch also provides a feed for earthworms, thereby providing one of the elements necessary to increase the earthworm population. Dr. Henry Hopp[11] studied earthworms in three clay soils, one with no cover, one with a fertilized sod cover, and one with a mulch cover. The results of this study are shown in Figure 4.3.

With no cover, earthworms had no effect on the relative infiltration rate of the soil. In this case, it is probable that the earthworms died from lack of food. When the soil was covered with a fertilized sod without earthworms added, the relative infiltration rate was increased by 200 percent over the uncovered soil. When earthworms were added, the combination of sod roots and earthworms increased the relative infiltration rate by 600 percent over the sod covering alone. When a mulch cover only was used, there was no increase in relative infiltration over uncovered soil. However, when earthworms were added to the mulch-covered soil, the relative infiltration was increased by an astounding 1600 percent. In the latter two cases, the organic matter furnished by the sod and mulch stimulated the earthworm population for both growth and activity and, subsequently, the greater network of burrows allowed the entry of more water at a faster rate. This data clearly indicated that the effects of earthworms alone on the relative infiltration is greater than either sod or mulch.

Figure 4.4 shows the effects of various cropping methods on erosion, water runoff, and earthworm population. While all of the facts are not known about the complete soil use, the data indicates that erosion, runoff, and earthworm population are fairly dependent on each other. Where the soil was continuously cropped to corn, there was no organic matter in the soil to furnish feed for earthworms; subsequently, there were no earthworms and, consequently, there was maximum runoff and erosion. Conversely, where the land was idle (a sod or weed cover is assumed) or in continuous meadow, the earthworm population thrived on the organic matter, there was no erosion, and water runoff was minimal.

It is clear from the data presented that earthworms are responsible for increasing the porosity, drainage, and moisture-holding capacity of a soil, provided, of course, there is organic matter for food. This, in turn, prevents erosion and water runoff.

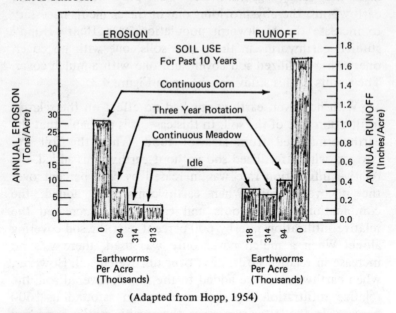

(Adapted from Hopp, 1954)

FIGURE 4.4—Effects of soil use on erosion, runoff, and earthworm population

5. Effects on Soil Aeration

Soil aeration, as described in Chapter 12, is one of the elements required for healthy and vigorous plant growth. Aeration affects the ability of a plant to exchange oxygen and carbon dioxide and is necessary to stimulate the microbial activity required to make nutrients available to plants. As stated in Chapter 12, a soil with good structure should contain 50-percent solid matter and 50-percent pore space, with pore space defined as the space between the solid-matter particles.

The aggregation of soil particles, the soil turnover and the burrowing effects of earthworms all affect soil aeration. As the soil is aggregated, or granulated into a "crumb" structure, the actual air space between crumbs increases and the earthworms' burrows tend to provide larger tunnels of air space between the crumbs. The constant mixing, or turnover, of soil by the earthworm maintains the entire topsoil layer in a loose and porous condition. The loosening of the soil caused by the aggregation of particles and the burrowing effects of earthworms is entirely different from that provided by cultivation. Cultivation provides an immediate loosening effect but, unless there is a large amount of organic matter in thessoil, this effect will last only until the first hard rain or watering. The soil will then start to compact again.

The efforts of the earthworm are continuous and its effects much longer lived since, the aggregates produced in earthworm castings do not readily dissolve water. Additionally, the sides of the earthworms' burrows are sealed with a similar water-stable mucous. It has been estimated that the burrows of earthworms occupy more than 5 percent of the total soil volume and that earthworms are directly responsible for increasing the soil air volume from 8 to 30 percent of the total soil volume.

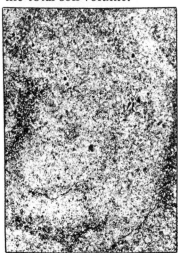
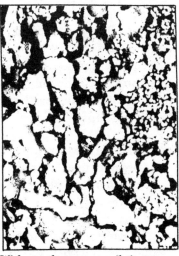

Without earthworms: soil heavy, packed, almost impervious to water. Cakes under hot sun; hard to cultivate; restricts plant and root growth.

With earthworms: soil is granulated, light, friable; absorbs water; is easy to cultivate; gives plant life a chance to grow normally.

(United States Department of Agriculture Photograph)

FIGURE 4.5—Effects of earthworms on the soil

6. Effects on Organic Matter Decomposition

The many factors involved in the decomposition of organic matter and the effects of the micro-organisms and small animals in the soil on the decomposition process are explained in Chapter 12 & 13. It is known that the passage of organic matter and mineral soil particles through the earthworm's digestive tract aids the process of organic matter decomposition, helps to solubilize plant nutrient elements from the insoluble soil minerals, and increases the structrual stability of the soil. However, the exact functions and the effects of earthworms on the decomposition of organic matter are not fully known.

It appears, from all available evidence, that earthworms perform several different functions which affect the decomposition of organic matter. It has been noted that there is always a large amount of undecayed organic matter on the surface of soils without earthworms whereas, when earthworms are present, this matter is fragmented and generally mixed throughout the soil. Generally, only the softest of organic materials are readily decomposed by microbial activity. The more resistant materials must first be broken into smaller pieces and the decomposition process started by the enzymes secreted from the intestinal tracts of the soil animals. The earthworm is probably one of the major contributors to this initial breakdown as well as to subsequent breakdowns of organic material, including the final process of decomposition, or humification. Sequentially, the role of the earthworm in organic matter decomposition is as follows:

a. The earthworm ingests fresh or partially decomposed organic matter from the soil surface, including the rougher, or more decay resistant, stems, roots and leaves.
b. The ingested organic matter, especially the tougher matter, is then fragmented, or broken into smaller pieces by the grinding action of the earthworm's gizzard.
c. The fragmented organic matter is mixed with enzymes in the earthworm's digestive tract. These enzymes, in turn, synthesize organic compounds which stimulate microbial activity in the earthworm's intestines. The microbial activity then starts the decomposition process.
d. Approximately 24 hours after ingestion, the earthworm excretes this fragmented organic matter in the form of casts on the soil surface and at various levels within the soil.
e. Since the organic matter is in smaller pieces, contains

some synthesized organic compounds and micro-organisms from the earthworm's digestive tract, microbial activity and, subsequently, decomposition are enhanced.

f. As the earthworm burrows through the soil, it reingests the fragmented and partially decomposed matter previously excreted as casts. This organic matter is again processed through the earthworm's digestive tract and excreted as casts in the soil, further enhancing the decomposition, or humification process.

The preceding process is probably repeated many times as the earthworm burrows through the soil. The end is that the original organic matter becomes a fine humus containing nutrients and minerals which are readily available for absorption by plant roots.

Earthworms can consume more organic matter from the soil surface than all of the other small soil animals together. However, earthworms will not consume certain types of organic matter when it first falls to the ground. In most cases, this matter contains tannic, citrus, or other acids or oils which must be leached out by natural processes before it will be consumed. The period of time over which the earthworms will break down the organic matter not only depends on its content, but also on the structural strength of the material and if it is dry or moist. Earthworms cannot consume any food in a dry state; it must be moist for easy assimilation. Moisture is also required for decomposition of the organic material by soil microorganisms.

Earthworms consume a large amount of food, for instance, mature individuals of *L. rubellus,* the common red worm, could ingest 0.56 to 0.7 ounces *dry weight* of manure per year. Based on this, Satchell[12] (1967) estimated that the 2.7 to 3.3 tons of manure per acre per year produced by dairy cattle is only one-fourth of the amount which could be consumed by a typical earthworm population. Satchell also estimated that earthworms could consume the annual leaf fall from a woodland in a temperate zone in about three months.

The preceding shows that if proper soil management practices were used on farms, wastelands, forests, etc., earthworms could consume and help decompose all of the organic waste materials generated by nature, man, and animals.

7. Effects on Soil Nutrients

In the past, many researchers have discounted the idea that earthworms could chemically change the nutrients and minerals in the organic matter that they ingest so that these are in a form which is readily available to plants. However, most researchers now generally agree that earthworms castings and soils with earthworms contain more available phosphorus and molybdenum, exchangeable calcium, magnesium, and potassium, and have a higher base-exchange capacity than soils without earthworms. The earthworm, as it burrows and feeds, takes organic matter from the surface, digestively processes it, and deposits the residue as casts (Chapter 8) throughout the soil in the plant root zone so that available nutrients in the casts may be readily used by the plants. However, the nutrients in the casts, both those which are available and those still in an unavailable form, depend on the content of the organic material and mineral soil—the earthworm cannot provide or make available nutrients which are not in the original material.

Experiments performed by Dr. Douglas Taff[13] proved fairly conclusively that earthworms do have the ability to modify mineral soil. This experiment was conducted in the following manner:

a. Containers were filled with 41 pounds of sandy loam soil.
b. Distilled water was added to prevent inducing additional elements.
c. Samples were taken from each container and frozen for later analysis.
d. A known quantity of *L. rubellus* (red worm) was added to part of the containers and the remainder of the containers were left without earthworms as controls.
e. After 40 days, samples were taken from all containers and analyzed for pH, positive ion exchange capacity, total nitrogen, nitrate and nitrite, ammonium, reserve phosphate, and available phosphate, calcium, potassium, aluminum, magnesium, manganese, iron, copper and molybdenum.
f. The final results were obtained by subtracting the nutrients which were added to the soil by the known number of dead earthworms from the analytical data.

g. The results showed an increase of 18.8 percent in available potassium and 68.2 percent in available manganese. There was no other difference in nutrients between the soil in the control containers and those with earthworms.
h. All earthworms were removed.
i. Both the control soil without earthworms and the soil which had been worked by earthworms were puddled to destroy the physical soil structure.
j. Both the control soil and the earthworm-worked soil were then seeded with a known quantity of rye grass.
k. Double-distilled water was used to maintain a constant moisture level and ensure that no additional minerals were added to the soil.
l. The rye grass was germinated and grown under lights.
m. After 14 days, the rye grass in the earthworm-worked soil was thicker, healthier looking, and taller than that in the control.
n. The leaves from both the control and earthworm worked soils were analyzed. There was no significant difference in nitrogen, magnesium, or calcium between the two samples. However, the phosphorus was 35.6-percent greater and the potassium 51.9-percent greater in the leaves grown in the earthworm worked soil than those grown in the control.
o. After four weeks, the difference was not as significant; the phosphorus content had stabilized at 0.42% and the potassium content was inversely proportional to plant size, that is, a larger plant has a larger volume.
p. After six weeks, the contents of all samples were equal and the potassium content had stabilized at 3.41-percent.

Dr. Taff's conclusion on this experiment was, "Based on these results, a person is able to argue that earthworms in actuality do have the ability to modify mineral soil. For example, if potassium, phosphorus, or other nutrients such as manganese are made more soluble by releasing them in their ionic forms, then this might partially explain the soil and leaf analysis data. Alternately, plant nutrients might become bound in organic chelates produced by the worm. This could make them available to a plant, but partially "invisible" to soil analysis. A third possibility exists in which earthworms produce organic compounds which are stimulatory to increased root growth. This could account for the faster growth rate and higher phosphorus and potassium content of the two-week-old rye."

"Plants which germinate and grow rapidly are usually healthier and more resistant to stress. If, as been shown with rye, you can increase a plant's rate of growth, its vigor and its chance of survival, then there appears to be an obvious agricultural reason for stimulating larger earthworm populations."

The fact that earthworm castings are higher in available phosphorus has been shown in experiments and studies by other researchers. Experiments at the Connecticut Agricultural Experiment Station have shown that earthworm casts contain approximately 5 times more nitrate, 7 times more available phosphorus, 3 times more exchangeable magnesium, 1.5 times more calcium, and 11 times more potassium than the surrounding soil.[14] A discussion on earthworm castings as well as analytical data is provided in Chapter 8.

The effect of the earthworm on the amount of available nitrogen is a subject of controversy among researchers; it is not as clear cut as the effects of earthworms on the availability of other nutrients. There is some statisitcal data and reports of an increase in nitrogen soils with earthworms. The organic matter which earthworms consume generally contains a large amount of nitrogen. Some of this nitrogen-bearing material is digested by the earthworm and extracted as protein for earthworm growth. Some of the absorbed nitrogen is later released as urea and ammonia by the earthworm and the remainder of the absorbed nitrogen and the nitrogen in the organic material which is not broken down are excreted as castings for later microbial decompositon. However, most researchers believe that this accounts for a small amount of the nitrogen in the soil. One study showed that, of the nonavailable nitrogen ingest by earthworms, only approximately 6 percent was excreted in a form which could be used by plants. Additionally, nitrogen-fixing bacteria do not seem to be affected by earthworms, although the numbers of *Azotobacter* appear to decrease when passing through earthworms. It is also thought that more nitrogen is excreted when the earthworm is semi-dormant and feeds less than when it is active and feeds more.

The more probably cause of a significant increase in available nitrogen in soils with earthworms is the decomposition of dead earthworms. Since 72 percent of the dry weight of an earthworm is protein (11.5 percent nitrogen), it is possible that the death of one large earthworm (smaller earthworms would yield less) releases up to 0.01 gram of nitrate. If this is correct, 1.5 million earthworms per acre could provide approximately 193 pounds of nitrate of soda per acre per year if it is assumed that the life span of an earthworm in the field is one year. This is over four times that required by most agricultural crops. Satchell (1967) estimated that 25 percent of the nitrogen added to the soil from decomposing earthworms was in the form of nitrate, 45 percent ammonia, 3 percent soluble organic compounds, and 27 percent miscellaneous microbial protein, seta, and cuticle.[15]

Thus, the combination of excreted nitrogen and that provided by the decomposition of dead earthworms, coupled with the earthworms' ability to make other nutrients available to plants shows that the earthworm does have a profound effect on the mineral properties of soil.

8. Effects on Carbon-to-Nitrogen Ratio

An explanation of the carbon-to-nitrogen ratio is provided in Chapter 14, where it was stated that the allowable carbon-to-nitrogen ratio for any organic material which is added to the soil is 30:1, or 30. It was also stated that the liberated nitrogen may be recycled by the soil microorganisms until the content decreases to a level where the carbon-to-nitrogen ratio was below 30 and that most plants could not assimilate mineral nitrogen unless the ratio was on the order of 20:1 or less. Since the carbon-to-nitrogen ratio plants cannot utilize any of the nitrogen in the organic matter until the ratio is below 20.

Earthworms help to lower the carbon-to-nitrogen ratio of fresh organic matter by consuming the matter, breaking it down, and using the carbon for energy during respiration. That is, the carbon is combined with oxygen and given off as carbon dioxide during respiration. While it is not cited in any of the available literature, it is assumed that, as with other animals, some carbon is also consumed (or "burned" or oxidated) by the earthworm's metabolism as energy during movement.

According to all statistical data, the earthworm has a very small effect on the carbon-to-nitrogen ratio. Of the total carbon consumed by micro-organisms and soil animals during the decomposition process, it is estimated that earthworms consume only between 4 and 12 percent by respiration. Consumption by other metabolic functions is unknown.

9. Effects on Soil Microbiology

It appears, from all available evidence, that earthworms, as they burrow through the soil, ingest large quantities of active and resting micro-organisms from the soil and deposit them elsewhere in the soil with their casts. There is generally no difference between the species of micro-organisms found in the digestive tract of the earthworm and in the surrounding soil. However, many researchers have found a significant difference in the quantities of micro-organisms in the earthworm's digestive tract and casts than in the surrounding soil with the numbers in the earthworm casts reported as three-to-five times greater than the surrounding soil. It has also been reported that there were greater numbers of bacteria and actinomycetes in the earthworm's digestive tract than in the surrounding soil and that these numbers increase exponentially from the front to the rear of the digestive tract.

The depth at which the earthworm lives, burrows, and feeds also appears to affect the numbers of micro-organisms in the digestive tract and casts as compared to the surrounding soil. In deep-living species, these quantities are approximately equal. In species which live in the upper part of the soil and feed on organic matter on the surface, such as *L. rubellus* (red worm), there are up to ten times as many in the digestive tract and casts as in the surrounding soil, depending on the organic matter ingested and the activity of the earthworm. More actinomycetes and spore-forming bacteria, and less fluorescing bacteria were found in the digestive tract of the deep-living species than in the surrounding soil. However, there were more actinomycetes, fungi, butyric acid-forming bacteria, and cellulose-decomposing bacteria in the casts of both species than in the surrounding soil.

It appears that the number of micro-organisms in the earthworm's digestive tract and casts is somewhat proportional to the quantity and quality of organic matter in the soil in which the earthworm lives. It has also been noted that the microbial populations in the earthworm's digestive tract start to change when the contents are excreted as casts. Since, in most cases, these casts are high in partially decayed organic matter and nitrogen in the form of ammonia and urea, they appear to stimulate microbial growth for decomposition of this material. It has been estimated that the microbial population in casts doubles within the first week after excretion and, while the types of microbes fluctuate, no increase in total numbers was noted during the next three weeks. Some researchers have reported that the total microbial population starts to decrease from the time the casts are excreted; however, the method used to determine the numbers was measuring the oxygen consumed and this method does not count the numbers of micro-organisms which go into resting stage. It has also been found that the micro-organism population follows seasonal fluctuation, being greater in spring and summer than in fall and winter. This also corresponds to the environmental conditions necessary for decomposition of organic matter.

It has also been reported by Edwards and Lofty in Chapter 7 of their book that earthworms may produce antibiotic substance which inhibits the growth of certain fungi, saprophytic bacteria, and non-acidfast pathogenic micro-organisms. In one test, the growth of a certain fungus was inhibited as long as earthworms were present but growth started when the earthworms died or were removed. In another test, earthworms appear to have helped control apple scab caused by substances being released in the springtime from leaves which fell the previous fall. The earthworms removed the leaves from the surface, preventing at least a part of the infection. Another report indicates a strain of bacteria, *Serratia marcescens,* is destroyed in the earthworm's digestive tract. While this particular bacterium is not dangerous, other forms in the same family are and it is speculated that they are destroyed by the earthworm.

Microbial activity may also have an effect on soil particle aggregation. As previously stated, this increases the water stability of earthworm casts in relation to the surrounding soil and may be attributed to the action of bacterial gums in the earthworm's digestive tract and fungi in the casts.

It has also been reported that certain protozoa are required in the organic matter on which the manure worm, *E. foetida,* feeds in order for individuals of the species to reach sexual maturity. It is speculated that this may be the reason why this species does not thrive in cultivated soils without a high organic matter content.

In summary, it appears that earthworms have a profound influence on the distribution and population of micro-organisms in the soil and stimulate microbe population by partial decomposition of organic matter in their digestive tracts. This partially decomposed organic matter is then excreted as casts containing a large quantity of nitrogenous material which further stimulated microbial activity and populations to complete the decomposition process. Earthworms also appear to be important in controlling certain adverse, or disease causing, bacteria, by producing antibiotic substances. Studies around the world, including one from Africa,[16] prove that earthworms can improve a soil, both structurally and chemically for the better, by their actions.

★ ★ ★

REFERENCES

1. Darwin, Charles, (1976) **DARWIN ON EARTHWORMS**, (original pub. 1881) Bookworm Publishing Co., Ontario, Cal., pp. 23 & 24.
2. **WHAT EVERY GARDENER SHOULD KNOW ABOUT EARTHWORMS**, op cit., p. 24.
3. **BIOLOGY OF EARTHWORMS**, op cit., p. 136.
4. Ibid. p. 137.
5. Ibid. p. 138.
6. Ibid. p. 139.
7. **DARWIN ON EARTHWORMS**, op cit., pp. 60 & 61.
8. **BIOLOGY OF EARTHWORMS**, op cit., p. 138.
9. **EARTHWORMS FOR ECOLOGY AND PROFIT, VOL. I.,** p. 105.
10. Ibid., p. 108.
11. **WHAT EVERY GARDENER SHOULD KNOW ABOUT EARTHWORMS**, pp. 16-17.
12. **BIOLOGY OF EARTHWORMS**, op cit., pp. 145 & 146.
13. Dr. Taff of Vermont, for Garden Way Publishing Co.
14. Lunt, H. A. and Jacobson, G. M. (1944). The chemical composition of earthworm casts. *Soil Science*, Vol. 58, No. 5.
15. **BIOLOGY OF EARTHWORMS**, op cit., p. 148.
16. Tran-Vinh-An (1973) L'action des vers de terre, genre *Hyperiodrilus africanus*, sur quelques proprietes pedologiques d'un sol sablonneux de la region de Kinshasa (Zaire), Cah. **ORSTOM**, Vol. XI., no. 3/4, pp. 249.

CHAPTER FIVE

TREAT EARTHWORMS KINDLY

As we have seen, earthworms do quite a lot for soils, including the soil in your garden. They improve soil structure and moisture absorption and holding capacity. They aerate the soil with their tunnels, providing room for growing roots. They transform and distribute nutrients the plants need.

Worms also help to neutralize soil pH balance, which can be an important factor for many plants. Earthworms have glands, called "calciferous glands" which secrete calcium. This calcium makes the soil more alkaline or "basic" than it would otherwise be. On the other hand, some researchers have found that earthworms can also help to correct an over-alkaline condition; although how this is accomplished is not yet clear.

A. What You Can Do

But what can you do for your worms, in exchange for all they do for your garden?

1. Feeding

One of the most important things you can do for your worms is to feed them. This means keeping plenty of organic matter in or on your soil and around your plants at all times. Animal manures, particularly cow, goat, rabbit, and the like, are excellent for this purpose. However, these must be used in moderation around the more sensitive plants, because some of these manures contain too much nitrogen, which can "burn" plant roots. This is particulary true of manure from chickens, and other poultry,which can contain 4% nitrogen or even more, too much for most plants unless the material is used very sparingly.

Compost made from biodegradable wastes such as old newspapers, table scraps, lawn & garden clippings, or even old rags, is another excellent source of organic nutrition for your worms, and for your plants. Earthworms can play a very useful role in your compost pile; a subject we'll discuss in Chapter 5.

GRAPHIC 5.1–(See Chapter 11)

But feeding worms is only part of the job. They also require moisture. Of course, you should water your garden regularly. For best results, you must set a definite schedule for watering; depending on the season and the weather. Obviously, you will water more often during the summer than in the early spring when the rains come frequently.

2. Watering

The object of a good watering program is to keep the soil moist but not soggy from the surface down to two feet or more below, depending on the type of plants you have. This can best be done by watering deeply rather than often. Frequent light sprinklings are not the best way to water. It's far better to water for an hour or two every other day, for example, than to water for 15 minutes three or four times during the same day. Frequent light watering goes not penetrate more than a very few inches beneath the surface. Thus, worms below this depth will dry out, and the plants themselves will develop very shallow root systems which are unable to provide all of the moisture and nutrition the plants need to reach their full growth.

The mulch or compost on the surface of your soil should also be watered, as the worms cannot eat dry food. Worms ingest their food by sucking it in through their toothless mouths, and it must be soft and moist in order to pass through. As most good gardeners know by now, another advantage of using a mulch is that it will help prevent too rapid evaporation of water from the soil. Thus, the water goes down where it should, instead of up into the air.

3. Proper pH

As mentioned, earthworms will help to maintain a proper pH balance, near neutral in your soil. But if your soil is too acid or alkaline to begin with, the worms will not be able to do the whole job by themselves.. In extremely acid or alkaline conditions, the worms themselves will die.

To prevent this problem, you should first buy a pH kit at a garden supply store, feed store, or worm farm. These are rather inexpensive and the directions for using them are easy to follow. You should use the kit to check the pH of your soil every other month; or when adding fertilizer or organic matter, particularly manures, from outside your own yard.

If your soil is acid, with a pH reading below 6.5, you should add lime, calcium carbonate, or another alkalizing agent. If too alkaline, say above 7.5; you may want to mix some peat moss into the soil. Peat moss tends to be on the acid side, although you need to check the pH of the peat moss too, since the amount of acid contained in different lots and brands will vary considerably, and you cannot always trust pH values, if any, given on the label of the bag you buy. If peat moss is unsuitable, check with your local Cooperative Extension Agent (County Agent), about the most suitable material to use in your area and where it can be obtained.

How much of any of these materials to use will depend, of course, on how far away from 7.0 (neutral) your pH is to start with, and on the type of plants you are growing. Plants will vary in their pH preferences, although most vegetables and common ornamental plants tend to do best when the soil is fairly close to neutral. But remember that *E. foetida* and *L. rubellus,* the most commonly sold species of worms for garden uses, will do best in soil or bedding which is no less than 6.8 and no more than 7.2; although they can survive a much wider range. *A. caliginosa,* on the other hand, which are widely used for soil improvement in Australia and New Zealand, are reported to favor more acid soils. The same is known to be true of the Canadian or Native nightcrawler, *L. terrestris,* which favors an environment around 6.5 pH.

The use of worms for garden soil improvement is frequently associated with the so-called "organic" style of gardening, which also calls for avoiding chemical fertilizers or pesticides. But the fact is that neither nitrogen fertilizers nor several common pesticides, properly and carefully used, are harmful to earthworms. Many studies have shown that the addition of some nitrogen fertilizers in proper amounts will actually increase the number of worms in the soil. We will discuss both fertilizers and pesticides in detail in following chapters.

4. Tillage

One practice which you should avoid is a too frequent disturbance of the soil through plowing or hoeing. Remember that the earthworms in your soil will be doing much of this work for you, and they will do better if not distrubed too often. Many gardeners are discovering, much to their surprise, their plants will actually do better with some weeds around them than they will if the soil is constantly being scraped and pulled, and then left almost entirely bare.

Of course, you must remove weeds that are growing too close to the base of your vegetables or flowers, and you do want your garden to present an attractive appearance. But weeds between your crop rows will also aid in soil moisture retention, and they make excellent additions to your compost pile. Or they can be turned directly into the soil as green manure.

If all of this advice about caring for earthworms in your garden sounds identical to advice about caring for your plants themselves, it should not be surprising. Nature, after all, is a single system; incredibly complex and beautiful to contemplate, but with an underlying unity at its heart. Life begets life, and conditions which favor life in the vegetable kingdom will also, as a rule, be the same conditions which are most favorable to life in the aminal kingdom, from earthworms to human beings.

B. WHERE TO BUY WORMS FOR YOUR GARDEN

There are a wide variety of places today where earthworms may be purchased for home gardens. When most people think of worms, they also think of fishing bait; but a bait store is not usually the best place to purchase worms for gardening. For one thing, bait shop prices reflect packaging and handling that's intended to benefit the fisherman, not the gardener, and so are often higher than prices to be found elsewhere. Another problem is that worms in bait shops are sold by the dozen or in cups of fifty, while gardeners need them by the thousands.

Worms are now available by mail from large mail-order nursery suppliers, who are offering earthworms for gardeners through their mail order catalogs and special mailings to customers. The full list of these companies would read like a Who's Who in American horticulture, including several seed companies and some general merchandise mail order houses as well.

Fishing and gardening magazines carry classified columns filled with advertisements for mail order sales of earthworms. While many of these ads are placed by well established vermiculture firms, others are sometimes inserted by small, newly established firms or individuals. In the latter case, some of these advertisers are not sufficiently knowledgeable about packaging requirements for successful long distance mailing of their product; and some may go out of business before they ship your order. So it does pay to be cautious in answering such ads, unless you are familiar with the individual company with which you will be dealing.

PHOTOGRAPH 5.2—With proper packing, live worms travel by air or rail quite safely.

Many local nurseries and garden departments of major stores now offer earthworms for sale. Unless they have refrigerated storage facilities on the premises, these stores will not generally have worms for sale over the counter. However, they will have an order form available. After you fill out the form and pay for the worms; the store will arrange to have your worms delivered by United Parcel Service or through the mail within a few days. North American Bait Farms, one of the nation's largest vermiculture marketing organizations, has established such programs with many local outlets.

One of the best places to buy worms for your garden is at a local worm farm. There you can judge for yourself the quality and count of the worms you are buying. You can ask questions about proper care for them in your garden, and judge the general quality of the company with which you're doing business.

C. ADDING WORMS TO YOUR GARDEN

The first step in adding earthworms to garden soils is to make certain that the soil itself is properly prepared to receive them. This means soil that is moist, with generous quantities of organic matter mixed in. The pH should be checked and corrected if necessary. The soil should be free of most pesticide residues, particularly organochlorines which are almost invariably and immediately fatal to earthworms. The soil should also be completely loose down to at least two inches depth, so that the worms can get below the surface quickly as they are added. They will gradually go deeper after that.

The number of earthworms to add will depend on the species you select. The species selection in turn will depend on your climate and on which species are most available in your area, and on the varying prices charged for them. Canadian nightcrawlers, (*L. terrestris*) will do well in colder areas of the country. Africans (*L. Africanus*) are only suitable in the warmer climates of the deep South and the desert areas of the Southwest where the ground never freezes. Redworms, (*L. Rubellus*) on the other hand, will survive and multiply almost anywhere, given proper protection with insulating mulches. Redworms are also generally much less expensive to purchase than either Canadian or African nightcrawlers. They are also more prolific than the other species mentioned, although they are smaller in size, and you will need more of them per square foot to maintain top growing conditions in the soil.

If you are using nightcrawlers, either Canadian or African, you will need about 12 worms per square foot of planted area. With Redworms, you should add about 25 worms per square foot. Another species which has shown itself to have outstanding soil improvement abilities, *A. caliginosa,* or field worm is unfortunately not generally available in the United States from commercial sources as this is being written.

When To Add Worms

The best time of day to add worms to your soil is early in the morning, while it is still cool. Mornings are better than evenings for this purpose, because the light will increase for several hours, driving the worms, which are photophobic (light-haters) deeper into the ground. By the time 12 or 14 hours have elapsed and darkness comes again, the worms will be well settled in. But if you add them in the evening, when light is rapidly fading, they may rise to the surface and be long gone by morning.

Take the worms you have obtained along with a bucket of compost or fresh cow manure obtained from a dairy (not from a nursery) to your inoculation points. Spread the worms across the area where you want them to work on top of the loosened soil. They will immediately begin to dive below the surface. As soon as they have disappeared, add a layer of compost about one or two inches thick over the surface where you placed the worms. After you have completed covering your garden area with the worms and compost, water thoroughly but not so much that the ground becomes muddy or waterlogged.

If you want to apply earthworms to a garden or planted area already growing, gently loosen the soil around the plants, and spread the worms over the loosened soil, and wait for them to penetrate. For medium sized plants, use about 50 worms (*L. rubellus*) around the base of each. For trees, or very large shrubs, you will need about 200 worms (*L. rubellus*), about two large handfuls. After the worms have penetrated, add a layer of compost about 1½ inches deep and water.

After you have added your worms, you should care for them in the same way as you would for native worms in your garden as previously explained. While you won't see them much, unless you go digging down into the soil, you will notice the difference they make in the production of your fruits, vegetables, and flowers. For the healthiest plants are generally found in soils with a numerous and healthy earthworm population.

D. PROTECTING YOUR WORMS

Keeping a plentiful covering of mulch over the top of your soil and around your trees and shrubs, is the best and most practical method of keeping your worms happy, winter or summer, rain or shine. The mulch, as noted previously, is a primary food source for surface feeding species, which all of the types we've mentioned by name are. But mulch, kept about one inch deep in summer and three inches deep in winter, is also an excellent insulating material. Ground covered by mulch stays cooler during the summer and warmer during the winter. This is particularly important in summer in desert areas, where temperatures go above 100 degrees F.; and in areas subject to ground frost during the winter months. Also the mulch acts to increase the moisture retention capacity of your garden, so your water is used more efficiently.

If your ground freezes during the winter, the population of such species as *L. rubellus* (redworms) can be significantly reduced, even with a mulch cover. However, you will not lose all of your worms in such cases, because many of them will work down below the frost line and lie quietly there until Spring. In addition, the capsules laid by these worms in the late Fall will remain dormant until the Spring thaw. Then they will hatch out to repopulate your garden. The same phenomenon of dormant capsules will occur if the soil becomes too dry during the summer. When the Fall rains come, out come the worms.

With proper care and attention, your earthworms will multiply and flourish as they increase your crop yields.

★ ★ ★

CHAPTER SIX

YOUR EARTHWORMS AND FARM RESEARCH

Charles Darwin, in the concluding paragraph of his famous work on earthworms, wrote: "The plow is one the most ancient and most valuable of man's inventions; but long before he existed that land was in fact regularly ploughed, and still continues to be thus ploughed by earthworms."[1] But, as we saw in Chapter 2, the earthworm is much more valuable to the soil than a simple plow. Earthworms are also significant contributors to the supply of nutrients in the soil and they promote the growth of vital bacteria both in their own intestines and in their castings. For a more complete review of this latter point, the reader is referred to BIOLOGY OF EARTHWORMS by Edwards & Lofty, Chapter 7, *Earthworms and Micro-organisms.*

A. AVAILABLE RESEARCH

From all that we have seen so far, we should expect that earthworm populations will be positively associated with increased production of many kinds of crops, and of livestock in pasture. Therefore, we shall examine the available research conducted in the field for various types of farming operations.

1. Pastures

The leading center of research on earthworm effects on pastures used for grazing livestock has been for many years located in New Zealand. There Drs. Stockdill and Cossens, researchers for the New Zealand Ministry of Agriculture & Fisheries, have studied the effects of earthworms on sheep pastures. Their research has been recognized as having considerable economic importance in New Zealand, since exports of wool and mutton have been mainstays of New Zealand's agriculture for more than a century.

Much of this research has already been mentioned in earlier chapters. However, in an article published in the New Zealand Journal of Agriculture in 1969[2], D. K. Crump, a farm economist, estimated that for every dollar a sheep

Soil Profiles from Australia

PHOTOGRAPH 6.1—With no earthworms, the soil is packed and dry.

PHOTOGRAPH 6.2—With earthworms, soil matter or organic matter is mixed throughout, creating a better soil structure.

(Both photographs Courtesy of W. T. GORDON)

farmer spent on adding earthworms to his pastures, he would realize a return of $3.34 per year. It should be noted that these figures are for New Zealand currency in 1969; however, the principle of the ratio of investment to increased profit remains the same today.

Crump went on to say, "If a farmer waits for 30 years or more it is possible that earthworms will eventually migrate into the pasture from areas around shelter belts or the homestead, although the waiting period could well be longer than this. The farmer who does not establish earthworms sacrifices a net return of at least (N.Z.) $4.70 an acre each year for at least 20 of the 30 or more years that he waits for earthworms to establish of their own accord."

Finally, in an article from the same publication[3] published in January, 1976, G. A. Martin and Dr. Stockdill report on the development of a machine to assist in the process of earthworm inoculation. Giving their reasons for undertaking this development they said: "The presence of earthworms can lift pasture production by between 25 and 30 percent, and can raise carrying capacity by 2.5 stock units per hectare." (See Chapter 11.) In other words, pasture with worms can support more sheep per acre than the same pasture without worms. Surely this is a finding of immense potential importance not only to sheep farmers, but also to cattle ranchers, horse breeders, and wildlife managers.

2. Orchards

Research on earthworms in fruit orchards is scarce, and the results reported are somewhat contradictory. Thomas Barrett, in his book **HARNESSING THE EARTHWORM**[4], devotes an entire chapter to recounting the experience of Frank Hinckley who managed several acres of orange groves by means of "earthworm tillage" for several decades with good results. While the Hinckley groves are filled with bright promise, the methods used and the results obtained were apparently not subjected to controlled studies. BUT as this direct quotation will prove, his crop yield doubled after the addition of earthworms.

"On one of my ten-acre groves, Hanford Loam, I discontinued all cultivation about eighteen years ago. At that time the twenty-eight-year-old trees appeared to have reached their limit as to size and production, about 300 boxes per acre per year.

The first year after changing my cultural method to one of non-cultivation, I noticed a great difference in water penetration. Plough sole was eliminated, the trees started growing and they have continued to do so ever since, until they now are large, fine trees, and my production average for the last fifteen years has been about 630 boxes per acre per year."

Hinckley added only soluble commercial fertilizers, either calcium nitrate or sulphate of ammonia. He gradually reduced this fertilizer to a dosage of 1 1/3 lb. of actual nitrogen per tree per year. He found the worms tilled his orchard, so he had only to hand rake the surface weeds once a month. The burrowing of the earthworms allowed Hinckley to use less water but for his trees to absorb more water. Earthworms saved him from spending thousands of dollars on farm machinery for tilling.

During the last decade, experiments have been conducted in Holland by J. A. van Rhee, to study the effects of earthworms on the productivity of apple orchards. Final results of these experiments have not yet been published; and preliminary results, while positive in some respects, are inconclusive. Reporting to the FOURTH International Colloqium of the Zoology Committee of the International Society of Soil Sciences,[5] Dr. van Rhee stated that the earthworms used (*A. caliginosa*) had a positive effect on soil structure, aeration, and root development. However, differences in fruit yield between trees with and without earthworms were not significant in the two years reported. As Dr. van Rhee noted, "there is a high year-to-year variability of fruit yields occurring in these climatic regions." Therefore, studies averaging production results from many years will be necessary to answer this question definitely.

In general, it must be said that the use of earthworms to promote the growth of various kinds of fruit bearing trees, is a subject which deserves a great deal more research activity than it has received so far. It is known from many studies that earthworms perform a number of functions in tree growing areas that are helpful in maintaining soil fertility: such as incorporating leaf litter into the soil and assisting in its transformation into humus, aerating the soil, stabilizing soil structure, increasing soil water absorption rates and holding capacity, etc. Several studies have shown that earthworms are responsible for each of these activities in fruit bearing orchards, particularly apple orchards. However, only controlled studies over an extended period, perhaps 10 years or so, can tell us definitely how much of a difference earthworms can make in orchard yields. Meanwhile, however, orchard and grove managers interested in increasing their yields by organic and biological methods, may want to inoculate with earthworms even before such studies are completed. They can already do so with confidence that the bulk of the studies done so far all point in the direction of positive effects, even though final answers have not yet been reported.

3. Field Crops

Many of the studies cited in Chapter 1 of this part demonstrated increased yields of various types of field and row crops from the addition of earthworms. Among the crops for which increases were demonstrated were barley, beans, clover, corn, millet, soybeans, hay and wheat. However, most of these experiments were conducted in more or less artificial settings in the laboratory rather than in fields actually being used in production. While results from such experiments can be highly suggestive, only results from extended field trials will be considered definitive by most farmers. In the light of present evidence, such trials are more than justified and should be a high priority on the part of the U.S. Dept. of Agriculture; Agricultural Research and Extension Services.

Among the justifications for such trials, the work of Dr. O. Atlavinyte of the Institute of Zoology and Parasitology of the Lithuanian Academy of Sciences in the Soviet Union must be counted very strongly. In a study co-authored with C. Pociene and published in *Pedobiologia* in 1973[6] Dr. Atlavinyte found that earthworms decreased numbers of algae in the soil under oat crops. But what was most significant from our standpoint is reported as a byproduct of their main research. As they put it: "It is to be noted that the effect of different numbers of earthworms (*A. caliginosa*) is also positive on the oat crops both in the vegetative pots and in field experiments. Oat crops increase by 50-20%."

In the same year, Dr. Atlavinyte published a second paper in the same scientific journal[7] co-authored with J. Vanagas. In this second paper, Drs. Atlavinyte and Vanagas discuss the role of earthworms in the accumulation and distribution of phosphate phosphorus (P_2O_5) and Potassium Oxide (K_2O) in soils under barley, bare soil, soil fertilized with straw and soil fertilized with mineral fertilizers. Following is a summary of their findings:

> "The number of earthworms and the season and duration of earthworm activities are of importance for the accumulation of mobile P_2O_5 and K_2O. The effect of earthworms on the amount of mobile P in the soil with plants, without plants, soil fertilized by straw, and soil fertilized by mineral fertilizers is always positive and differs only in its degree of intensity. There is a higher effect of earthworms on the accumulation of P_2O_5 in the soil without plants and that fertilized by straw. The least effect was observed in the soil with plants (barley). In soil under plants, faster growing plants such as barley use up P_2O_5 and K_2O accumulated there due to the activity of earthworms in proportion to the numbers of earthworms present."

In an earlier study, published in *Pedobiologia* in 1969, Dr. Atlavinyte and Dr. Daciulyte found that several species of earthworms (*E. rosea, A. caliginosa, L. rubellus,* and *L. terrestris*) all had positive effects on the accumulation of vitamin B_{12} in soils. They reported that B_{12} increased from 2 to 12 times depending on the number of earthworms, the period of their activity, and the humidity and temperature of the soil. Measurements were taken periodically from 4 to 12 months after the earthworms were added to the soils being tested.[8]

a. Problems in current research

Aside from the simple lack of adequate funding and interest in earthworm research, particularly in the United States, there are a number of other considerations which must be mentioned in connection with the use of earthworms to increase agricultural production, so that the reader will have a complete picture and not be misled.

The most important of these considerations is the matter of interspecific differences in earthworm activity associated with plants. It must be noted that not all species of earthworms will produce the same effects on soils and plants. Nor, as noted by Drs. Hopp & Slater[9] will the same species of worm produce the same degree of effects on different species of plants.

The earthworm species most frequently reported in connection with plant studies is *Allolobophora caliginosa,* although there are a number of other species which have also been observed to produce some, if not all, of the same effects, though not necessarily to the same degree as *A. caliginosa.* This worm is not generally found as the dominant species in commercial production, although it does occur in some commercial beds.[10] Often called a field worm, *A. caliginosa* is our most abundant species in the United States. In New Zealand, grass sods rich in *A. caliginosa* are removed in strips and replanted in pastures that lack worms. Soil scientists have found these earthworms can survive and multiply and spread outwards safely from this natural type of inoculation.

On the other hand, B. M. Gerard reported to a 1962 meeting of soil scientists, [11] that "no single species of earthworm is responsible for all of the contributions often attributed to 'earthworms'; the total contribution results from the activities of several species in the field. It is probable that each group of species having the same mode of life, or even each individual species has a particular association with other soil organisms. When more is known about these relationships it will be possible to show which species of *Lumbricidae* play important roles in soil fertility and also why."

Other species are questionable at this stage of little completed research. Grant[12] reported in 1955 that *E. foetida* would not survive in agricultural soils, and Stockdill[13] is of the same opinion insofar as introduction of *E. foetida* to pastures is concerned. Yet there is nothing in the literature to correlate specific characteristics of this species with inability to produce beneficial effects on plant yields, and many organic gardeners have informally reported favorable results obtained through the addition of commercially raised earthworms to their garden plots. This may be due to their addition of much compost or manures to feed the worms. Given the scarcity of "hard scientific" evidence on this point and the abundance of "soft" or "organic" testimony in favor of commerically produced worms, we believe this question must be treated as still open to further research and debate, at least until the "hard scientists" can satisfactorily account for the effects attributed to commercially produced worms by organic gardeners, and prove conclusively that these effects are in fact produced by other means.

B. A PROGRAM OF AGRICULTURAL RESEARCH

We believe that the following steps should be undertaken as a nationally coordinated program of research under the sponsorship of the Agricultural Research Service of the U.S.D.A. and/or the National Science Foundation.

1. Taxonomic Training and Clarification

In this year of 1976, the United States does not have a single qualified researcher who has a doctorate degree in Oligichate Annelidic taxonomy.[14] There is not one in active practice. A taxonomer is a scientific classifier. Very simply, we do not have a single person scientifically trained at a university to tell a *E. foetida* from a *L. terrestris*. Of course, worm producers or vermiculturists, have gained such necessary knowledge from experience or reading in their field of work. Therefore, the recruitment and training of persons qualified in this specialty must be a matter of the very highest priority and prospective students in this field should be subsidized in their studies in the U.S., Canada, England or elsewhere.

While the training of such specialists and field workers is essential, and now something of a scientific "emergency", it is also important to refine the current systems of taxonomy, by working with such distinguished scientists as Dr. J. W. Reynolds of New Brunswick and Donald Schwert of the University of Waterloo. Current systems of taxonomy are based entirely on physiological characteristics, either external or internal, which may or may not be significant in terms of the usefulness of the particular species to Man. The problem of many unrelated names being applied to the same species of earthworm has been pointed out by a number of writers, including Gates[15], Schwert[16], and Reynolds[17]. It is also clear that a number of studies have been published in which the species identification is simply wrong: for example, an otherwise excellent paper by Drs. Fosgate & Babb of the University of Georgia[18] which will be reprinted in a later chapter. The seemingly endless proliferation of species reported in the taxonomic literature gives rise to the suspicion that, at least, some of these reports may be spurious, particularly in light of the extremely limited body of true taxonomic expertise actually available today in the world.

What is needed is a standard reference, providing both verbal and photographic identification of those species which are thought to be of some significance, in commercial production and their applications today. Such a reference, imposing standard nomenclature (names) for all workers in the field around the world, and providing a ready guide for the naturalist, ecologist, and soil scientist, would be of immeasurable benefit to the entire science of vermology, not to mention the fields of agronomy and soil biology. Once published, such a reference should be disseminated as widely and as quickly as possible.

2. Test Planting Research

The numerous studies previously cited which show significant positive effects of earthworm populations on yields of a variety of commercial crops, provide ample justification for a coordinated program of test planting research; particularly when we are so intensely and justifiably concerned to discover methods of increasing agricultural production without depending on increasingly scarce fossil resources (such as natural gas for fertilizers) and without further damaging our natural environment.

The aim of these test planting studies should be to correlate activities of particular species or groups of species of worms with particular species of plants, under reproducible conditions. It may well be anticipated that such a program would lead to the discovery that some plants are most favorably affected by one species or group of earthworms, while other kinds of plants are aided most by other species when the feed and other factors are favorable. Climate, moisture, temperature, and soil type, undoubtedly are all involved in these relationships and it is essential that we discover exactly how these relationships occur in nature, and how they may be modified for maximum agricultural yields. Such research will take years, perhaps decades, to produce meaningful results, which is all the more reason it should be initiated as quickly as possible. As the age of chemically based agriculture is swiftly passing, and unless the technological base for a biologically based agriculture to replace it is organized quickly, we face grim prospects indeed.

3. Vermiculture Technology Development

The vermiculture industry has made enormous strides during the past several years in developing new methods of rearing, harvesting, shipping and marketing earthworms. Now we must refine and improve on what has been accomplished so far. Basic research into the life cycles of commericially produced species of earthworms is badly needed. Dr. Virginia Vail of the Tall Timbers Research Station in Florida has already provided us with an excellent example of the sort of study involved here, by her work on egg capsules from *E. foetida*[19]. But much more of this kind of study is required.

FIGURE 6.3—Earthworm Capsules

We also need both basic and applied research, particularly in the area of earthworm genetics; which could lead to new species of worms: species which would grow more rapidly to full size, and which might combine the soil enhancing capacities of *A. caliginosa*, with the producibility of *E. foetida*. This is particularly important, since *A. caliginosa* cannot be produced commercially by present techniques,[20] and therefore can play only a limited role in large scale inoculation programs, due to the expense and uncertainty of harvesting and transporting such worms from a wild state.

Also, research in the techniques of earthworm application in the solution of agricultural problems should be a high priority. Optimum methods of inoculation, for example, in various types of gardens and fields, need to be developed and disseminated.

4. Professional Education

Research results are only useful if they are disseminated through education to the professional workers who can make use of them. Today there are enormous gaps in both professional and popular understanding about earthworms, which must be filled by aggressive education programs carried on by the Cooperative Extension Service of the U.S.D.A. and other agencies.

Many businesses also need to educate their workers about earthworms, their value, and the techniques for handling and using them successfully. Airline freight workers are a prime example of a group which should receive this kind of education. Thousands of dollars worth of earthworms are lost each year due to careless and ignorant handling by airline freight personnel. In most cases, these losses could be easily prevented, if the workers involved understood and were willing to follow the simple procedures necessary to ship worms safely across the country.

Any industry today is dependent for its success on cooperation from other business groups, and the vermiculture industry is no exception to this rule. This is particularly important in the case of service industries such as transportation and finance, packaging and retail sales. Airlines, banks, department stores, nurseries, and packaging and machinery manufacturers can all increase their sales and profits substantially through involvement with vermiculture: by transporting, financing, selling, and packaging earthworms, and by making harvesting and other machines which can be used in the vermiculture industry itself.

Underlying these developments will be the growth of comprehensive and reliable statistical and other data relating to the vermiculture industry itself by the U.S. Dept. of Commerce and other government statistical reporting agencies. Today no one really knows how many earthworm farms there are, what the total value of sales is, the average value of sales, numbers of people employed full and part-time by this industry, etc. The development of such information must be a high priority by these agencies if the vermiculture industry itself is to grow and to make the contributions of which it is capable to agriculture and to our society as a whole.

5. You Can Help

A. Write to U.S.D.A. with a copy to your congressman and another to your Senator. In your letter, describe the needs that you see for additional research on earthworms, and tell why you believe such research will be of benefit to farmers and others in our country. Address your letter to:

>Director, Agricultural Research Service
>U.S.D.A., Washington, D.C. 20214.

B. Contact your local County Agent, for the State office of the Cooperative Extension Service. Ask what they are doing in this area, and what suggestions they have for local research activity. Send them a copy of this book.

C. Join with others who are interested in this area through your local unit of the American Farm Bureau Federation, better known as simply FARM BUREAU. If you are an earthworm grower, join the Farm Bureau and encourage other wormgrowers you know to do the same. When you have enough wormgrowers as members, (the number varies from state to state) ask for help in forming a Vermiculture Commodity Department at the local level, and a statewide Vermiculture Advisory Board. Through Farm Bureau, you will be able to press for additional research activity as group, and far more effectively than you can as an individual. Such units are already operating in a Farm Bureau in Arizona, California, Nevada, and other states. For additional information on how you can form a Vermiculture Commodity Department in your local Farm Bureau, write to:

>H. R. Heritage
>American Farm Bureau Federation
>220 Touhy Avenue
>Oak Park, Illinois

FARM BUREAU offers quite a number of benefits to its members, and can help you and your fellow vermiculturists in a large number of ways with everything from market research to cooperative purchasing of needed supplies. How much help you will receive will depend primarily on the number of members you can recruit for the vermiculture commodity department, and/or Vermiculture Advisory Board.

Do something today for worms, yourself and the world's growing and hungry population. Petition your politicians and Farm Bureau and even the public through private or public (letters-to-editors column) letters explaining the needs for vermiculture research.

★ ★ ★

REFERENCES

1. Darwin, C. 1976. **DARWIN ON EARTHWORMS**, pg. 148. Bookworm Pub. Co., Ontario, CA 91761.
2. Crump, D. K. 1969. Earthworms—a profitable investment. *N.Z. J. Ag.* 119 (2): 84-85.
3. Martin, G. A. & Stockdill, S. M. J. 1976. Machine speeds the job of earthworm introduction, *N.Z.J. Ag.* 126 (1) 6-7.
4. Barrett, T. 1976. **HARNESSING THE EARTHWORM**, page 55-62. Bookworm Pub. Co., P.O. Box 655, Ontario, CA 91761.
5. van Rhee, J. A.; 1971. Productivity of orchards and earthworms., *Proceedings*, 4th Coll. Zool. Comm. Intl. Soc. Soil Sci. pg. 99-107.
6. Atlavinyte, O.; Pociene, C. 1973. The effect of earthworms and their activity on the amount of algae in the soil. *Pedobiologia* 13 (6) 445-455.
7. see Chap. 2, note 13.
8. Atlavinyte, O.; Daciulyte. 1969. *Pedobiologia* 9:165-170.
9. Hopp, H.; Slater, C. S. 1949. The effect of earthworms on the productivity of agricultural soil. In **THE CHALLENGE OF EARTHWORM RESEARCH**, R. Rodale, ed., 1961, Soil & Health Foundation, Emmaus Pa. pg. 67-83.
10. Causey, D. 1961. The earthworms of Arkansas. In **THE CHALLENGE OF EARTHWORM RESEARCH**, op. cit. pg. 43-56.
11. Gerard, B. M. 1963. The activities of some species of lumbricidae in pasture-land. In **SOIL ORGANISMS**, North-Holland Publishing Co., Amsterdam; pg. 49-54.
12. Grant, W. C. 1955. Studies on moisture relationships in earthworms. *Ecology*, 36 (3) 400-7, cited in **BIOLOGY OF EARTHWORMS**, op. cit.
13. Stockdill, S. M. J. 1976. Correspondence. Vermilogical Research Collection, Ontario Public Library, Ontario, Calif. 91761.
14. John W. Reynolds, formerly with the Tall Timbers Research Station in Tallahassee, Florida, returned to his native Canada in 1976. He was the only practicing Annelidic taxonomist in the U.S. specializing in oligichates. Dr. G. E. Gates, also formerly with Tall Timbers, is now retired. G. E. Gates of Zoology Dept., University of Maine, Orono, and John Warren Reynolds, Faculty of Forestry, University of New Brunswick, Fredericton, Canada, have both published much research on earthworms.
15. Gates, G. E. 1974. Contributions to a revision of the family Lumbricidae, *Bull. Tall Timbers Res. Sta.* No. 16: 9.
16. Schwert, D. E. 1975. References on E. foetida. Vermilogical Research Collection, Ontario Public Library.
17. Reynolds, J. W. 1974. The earthworms of Tennessee. *Bull. Tall Timbers Res. Sta.* No. 17, 133 pgs.
18. see Chapter 9.
19. Vail, V. A. 1974. Observations on the hatchlings of E. foetida & B. tumidus. *Bull. Tall Timbers Res. Sta.* No. 16: 1-8.

CHAPTER SEVEN

WORMS AND YOU TOGETHER

In this chapter we shall examine some of the effects that earthworm and human activities have on each other. We have already seen how some earthworm activities affect soils and plants. Now we will examine the effect that some agricultural practices have on earthworms; and the effect that some earthworm activities have of enhancing certain agricultural practices.

A. TILLAGE

Mechanical tillage of the soil is one of the oldest agricultural practices known to Man. Plowing has always been seen as an essential preliminary to planting. But recent research in agronomy both in the United States and in Germany is giving increased weight to the opinions of some "organic" farmers that mechanical tillage is simply another way of interfering with Nature, and that the less tillage the better.

In general, untilled land will contain more earthworms per sq. ft. than land which is regularly plowed,[1] although there are exceptions to this rule in the literature. Depth, method, and timing of tillage all affect earthworm populations. Some workers have postulated that tillage reduces earthworm populations by mechanical damage to the worms. While such damage does undoubtedly occur, it is unlikely that it would be a major factor in reducing a large worm population. Another theory holds that tillage which removes crop residues from the soil interferes with earthworm food supplies, and this hypothesis probably has considerable merit.[2]

However, work by Ehlers & Baeumer,[3] leads us to present another theory here. Ehlers & Baeumer showed that the plow "pan" (area of the soil in most direct contact with plow surface—smeared, compacted and flattened by the plow blade) interrupted earthworm burrows about 25-30 centimeters (8 to 10 inches) below the surface. It is possible that such a super-compacted area in the soil could "trap" large numbers of earthworms beneath it, preventing their access to the surface for feeding and cast evacuation. Such interference with their natural habits might very well lead to a substantial reduction in their numbers.

While one of the primary purposes of soil tillage is to increase permeability to air and water, research published in 1975 by Dr. Ehlers, indicates that for fields where large earthworm populations are present, such tillage may be not only unnecessary, but actually *counterproductive*. We believe Dr. Ehlers' paper will be of sufficient interest to our readers to justify its reprinting here by permission of the Williams & Wilkins Co., publishers of *SOIL SCIENCE*, where the article first appeared in 1975, Vol. 119, No. 3.

OBSERVATIONS ON EARTHWORM CHANNELS AND INFILTRATION ON TILLED AND UNTILLED LOESS SOIL

W. EHLERS
University of Goettingen, Germany[1]

Received for publication October 18, 1973

ABSTRACT

The number of earthworm channels ranging from 2 to 11 mm in diameter were counted to 80 cm depth in tilled and untilled grey-brown podzolic soil derived from loess. Number and percentage volume of earthworm channels in the A_p horizon approximately doubled during 4 years of no-tillage practice as compared to the tilled plot. Almost all the channels of the A_p horizon in the untilled plot had ports at the soil surface and were capable of taking in tension-free irrigation water and transmitting free water to a maximum depth of 180 cm where the channels end at the transition to unweathered loess. The channels in the A_p horizon of the tilled plot were not effective in water transmission. The same is true for a large number of channels in the subsoil of tilled and untilled plots, which could not contribute to water infiltration because they were not connected with the soil surface. The maximum infiltrability of conducting channels in the untilled soil was computed as more than 1 mm (1 liter per m^2) per minute, although the volume of these channels amounted to only 0.2 vol. %.

Limited water infiltration in soils is a severe problem in many regions, since it may hinder crop production and enhance soil erosion. Grey-brown podzolic soils derived from loess are widely distributed in Central Europe, and all show signs of poor infiltrability, especially when the soils are intensively cultivated. One of the reasons for slow water intake is the formation of a dense surface crust by clay-silt segregation at relatively low rain intensities. This aggregate lability again is caused by the high silt content, low organic matter content, and the absence of calcium carbonate. The surface crust acts like a seal, as the porosity and the number of large pores are decreased. McIntyre (1958) determined the saturated hydraulic

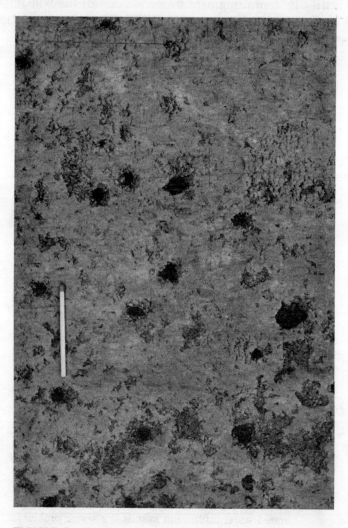

FIGURE 7.1—Horizontal cross section of soil in 50 cm depth with earthworm channels (B_t horizon).

conductivity (K_{sat}) of a crust formed by an Australian fine sandy loam as 5×10^{-7} cm/sec, whereas the conductivity of the soil below the crust was 10^{-3} cm/sec.

Another reason for slow water intake of grey-brown podzolic soils might be the formation of a traffic pan at the 20- to 25-cm depth with a very small porosity and a low content of pores $>30\mu$ (Ehlers 1973a). K_{sat} of the traffic pan in one of the silt loams in Goettingen was determined[2] as 8×10^{-4} cm/sec, whereas the uncompressed cultivated soil had a K_{sat} value of 5×10^{-3} cm/sec.

Unlike the conventional tilled loess soils, field plots of untilled loess soil in Goettingen never exhibited limited water infiltration. Zero-tillage (Baeumer and Bakermans 1973) became an important subject of research and in the United States a substantial method of corn cultivation after the development of nonselective herbicides with short residual effect. In the no-tillage system there is no longer any turning and loosening of the soil material. Plant residues are left on the soil surface, where they form a mulch cover. The seed is brought into the soil by direct drilling with a triple disc sowing machine.

After omitting soil tillage we recognized some characteristic changes in soil structure in the untilled plots. First, the structure in the upper 20-cm layer became more dense and porosity decreased as compared to the tilled plots (Ehlers 1973a, 1973b). The dense structure proved to be more stable, and in combination with the

FIGURE 7.2—Earthworm castings on untilled soil surface.

protective effect of the mulch cover, slaking of soil aggregates and formation of a dense soil surface crust ceased. Second, the very dense traffic pan loosened up after some years of no-tillage practice (Ehlers 1973a). Therefore, the main causes of limited water infiltration were less effective in the zero-tillage than in the conventional tillage system.

FIGURE 7.3—Exposure of earthworm channels down to 80 cm soil depth. The channel walls are marked with ultramarine blue color of the irrigation water.

Several investigators observed that the number and activity of earthworms increased when the soil was covered by mulch (Graff 1969; Teotia et al. 1950) and when tillage intensity was reduced (Becker and Meyer 1973; Schwerdtle 1969). In context with these observations we learned that Dixon and Peterson (1971) developed a channel-system concept describing the mode and intensity of water infiltration as a function of surface roughness and openness. The authors stressed the profound influence of large pores on water movement and showed that infiltration could be increased within a few months by undisturbed earthworm activity.

Therefore it was suggested that on the untilled land, water infiltration could have been enhanced in addition to the reasons stated above by continuous earthworm channels, which should connect the surface with the subsoil. In the present work we will report on the verification of this hypothesis. The results were obtained by counting earthworm channels in the tilled and untilled soil profile and proofing water infiltration through earthworm channels by use of colored irrigation water.

SOIL AND CLIMATE

Of soil and climate we state briefly the following properties: Soil type: Grey-brown podzolic soil (Typudalf). Parent material: Loess. Sequence of horizons in the profile: A_p (0-25 cm), A_1 (25-45 cm), B_t (45-110 cm), B_v (110-180 cm), C (>180 cm). Analysis data of A_p horizon: clay (>2 μ), 15 percent; silt (2-60 μ), 82 percent; organic C: 1.3 percent, pH: 6.7. Elevation: 160 m. Mean annual temperature: 8.2 °C; mean annual rainfall: 613 mm. Last tillage on the untilled plots: 1967. The investigations of which we will report were conducted in the summer of 1971.

PROCEDURE

Estimation of the volume of earthworm channels

A rectangular plot of 80 by 50 cm was investigated. Horizontal soil layers varying between 2 and 20 cm in thickness were removed successively up to 80 cm total soil depth with a spade. After a layer had been removed carefully, the number of earthworm channels was counted. They were classified into three size groups according to diameter: 2-5, 5-8, and 8-11 mm. At each depth the percentage area occupied by the worm holes was calculated by taking an average diameter for each size group. Later this percentage area was converted to percentage volume per unit thickness of soil layer.

Since the soil was investigated in dry periods, only a slight smearing of the soil occurred during removal of the layers, so that the holes were still identifiable. Sometimes, though, punching of the channels with a pencil or wire was necessary to get the hole free of soil for diameter estimation. Counting was difficult in the crumbly, tilled soil layer. The method allowed only the reliable identification and quantitative analysis of pores $\geqslant 2$ mm. The investigations were made in three replicates.

Demonstration of water infiltration through earthworm channels

Plots of undisturbed soil surface 80 by 50 cm were irrigated with 60 mm of colored water (ultramarine blue) for 5 hrs. On the next day the channels with blue walls could easily be exposed and identified in the profile by removal of vertical soil layers. The diameters of these channels were measured and the percentage volume calculated.

Determination of maximum cumulative water intake through earthworm channels

The upper uneven soil layer (2-cm thickness) of the untilled plot was removed to get free access of water to earthworm channels. Water was poured directly into the hole from a burette (small worm channels) or from a calibrated flask with a squirt (big worm channels) in such a way that the surface of the water was always to be seen in the port of the channel. The time for water intake was generally limited to 5 min.

RESULTS

Table 1 shows the number of earthworm channels per m^2 in the tilled and untilled plots. They are differentiated for soil depth and diameter. In the A_p horizon and in 30-cm soil depth the number of earthworm channels is substantially increased in the untilled soil as compared to the tilled soil. Largest numbers of channels were counted in 60-cm soil depths on both tilled and untilled soil. Figure 1 illustrates the frequency of channels in 50-cm soil depth.

Counting of channels and diameter estimation was impossible at the surface of the soil. The tilled soil surface

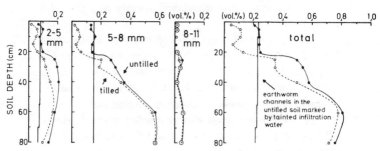

FIGURE 7.4—Percentage volume of earthworm channels in tilled and untilled soil for three different channel diameters (upper part) and for total channel number (lower part). Large circles with dot represent coincident observations for tilled and untilled soil.

was rather uneven and crumbly and the untilled soil surface was covered with straw. Part of this straw was concentrated by the earthworms in a star-shaped manner at the surface openings of the channels, where it was mixed with excrements (Fig. 2). The number of these casts was counted by Emanuel. He determined a mean of 55 casts per m^2 on the untilled soil surface. This quantity corresponds fairly well with the number of channels with a diameter >5 mm in the untilled A_p horizon (Table 1).

TABLE 1—Number of earthworm channels per m^2 in tilled and untilled soil

Depth (cm)	Tilled				Untilled			
	2-5	5-8	8-11	Total	2-5	5-8	8-11	Total
	(diameter in mm)				(diameter in mm)			
2	21	5	1	27	75	40	2	117
20	60	18	1	79	99	41	1	141
30	124	58	5	187	209	91	5	305
60	174	165	9	348	183	172	8	363

Figure 3 shows the calculated volume of the earthworm channels as percent of the total soil volume down to 80 cm depth (solid and broken line with dots and circles, respectively). The same figure shows the volume percentage of channels that had allowed infiltration of blue water into the untilled soil (nearly straight solid vertical line).

The exposure of these marked channels is shown in Fig. 4 down to 80 cm soil depth. Some of the deep-reaching channels were coated with color down to 180 cm depth, where these channels end at the transition to the C horizon (loess) containing calcium carbonate. In the tilled soil not a single blue channel could be detected. All the color was distributed in the tilled A_p horizon in a clusterlike, diffusive form.

Figure 5 represents the maximum cumulative water intake in 5 min. of channels with different diameters. In general the water-intake rates in milliliters per minute decreased slightly in time. Nevertheless, some of the 7- to 8-mm channels drained 2 to 4 liters when intake time was extended to half an hour.

DISCUSSION

Within 4 years of no-tillage practice (1967-1971) the number and volume percentage of earthworm channels nearly doubled in the A_p horizon (represented by 20-cm soil depth) as compared with the tilled plot (Table 1 and Fig. 3). By conventional tillage practice, earthworm channels are destroyed constantly in the upper soil layer. The quantity of channels in the tilled A_p horizon is the result of earthworm activity within one year. Altogether the volume percentage occupied by channels is rather low; it is less than 1 vol. % even in the B_t horizon, which had the highest density of channels in both tilled and untilled plots (Fig. 1). Here the main volume consists of channels 5 to 8 mm in diameter. Probably they are built up mainly by *Lumbricus terrestris* L. (Graff 1953; Schwerdtle 1969; Wilke 1962).

Unlike earthworm channels in the A_p horizon of the tilled plot, those of the untilled plot were capable of taking in irrigation water. Almost all of the channels reaching the untilled soil surface transmitted water deeply into the profile. This is shown by Fig. 3, where the percentage volume of channels, that transmitted blue water corresponds closely to the calculated percentage volume of the counted channels. While percolating down within the earthworm channels the water must have been in a tension-free state, because otherwise the ultramarine blue color would have been adsorbed by the soil matrix of the upper soil layer.

On the tilled plot it was observed that many of the channels in the A_p horizon were almost horizontally directed over short distances. Furthermore, most of them were blocked with loose soil aggregates. This soil material acts like a barrier for tension-free water. Reaching this barrier, free water undergoes suction and will be transported in the soil matrix due to an existing hydraulic gradient. This explains why none of the earthworm channels in the tilled plot was coated with blue color below the A_p horizon.

The results of the experiment show clearly that only those channels are good for water drainage that reach the soil surface and are not blocked by soil particles. The majority of earthworm channels in the subsoil cannot share in water infiltration.

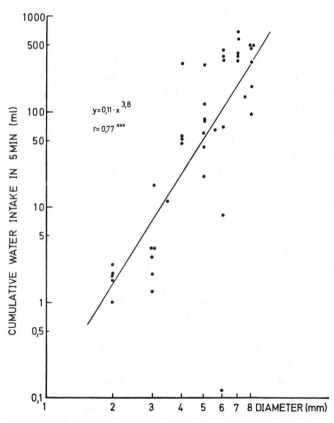

FIGURE 7.5—Maximum cumulative water intake by earthworm channels as related to channel diameter.

Until now, only a few indications of water infiltrations in macropores have been published (Dixon and Peterson 1971; Scharpenseel and Gewehr 1960; Williams and Allman 1969), and nothing is known about how much water can be transmitted in earthworm channels. Figure 5 shows that the cumulative water intake in 5 min. increases rapidly with channel diameter. Disregarding general water infiltration within small pores of the soil matrix, one can calculate a maximum infiltrability for the untilled soil, attributed to earthworm channeling only, by taking the average cumulative water intake in 5 min. for the 2-5 mm channels as 10 ml, for the 5—8-mm channels as 120 ml, and for the 8—11-mm channels as 300 ml. If one multiplies each by the number of channels per m^2 in 2 cm depth (Table 1), an infiltration rate of 0.15 mm (liters/m^2) per min is computed for the 2—5-mm channels, of 0.96 mm per min for the 5—8-channels, and of 0.12 mm per min for the 8—11-mm channels. Addition of these individual infiltration rates yields a total infiltrability of more than 1 mm per min, made possible just by conducting earthworm channels.

Water infiltration through earthworm channels can only be expected at high rain intensities, because at low intensities no tension-free water will exist at the soil surface and all the water will infiltrate through the soil matrix according to hydraulic potential gradients. One can regard the tubes as an additional drainage system within the soil matrix that becomes effective at high rain intensities.

REFERENCES

Baeumer, K. and W. A. P. Bakermans. 1973. Zero-tillage. Adv. Agron. 25: 77-123.

Becker, K.-W. and B. Meyer. 1973. Abbau, natuerliche Bodeninkorporation und Ertragswirkungen von Ernterueckstandsdecken (Stroh, Ruebenblatt) auf Ackerparabraunerden aus Loess. Goett, Bodenkundl. Ber. 26: 1-38.

Dixon, R. M. and A. E. Peterson. 1971. Water infiltration control: a channel system concept. Soil Sci. Soc. Am. Proc. 35: 968-973.

Ehlers, W. 1973a. Strukturzustand und zeitliche Aenderung der Wasser- und Luftgehalte waehrend einer Vegetationsperiode in unbearbeiteter und bearbeiteter Loess-Parabraunerde. Z. Acker Pflanzenbau. 137: 213-231.

Ehlers, W. 1973b. Gesamtporenvolumen und Porengroessenverteilung in unbearbeiteten und bearbeiteten Loessboeden. Z. Pflanzenernaehr. Bodenkd. 134: 193-207.

Graff, O. 1953. Die Regenwuermer Deutschlands. M. u. H., Schaper, Hannover.

Graff, O. 1969. Regenwurmtaetigkeit im Ackerboden unter verschiedenem Bedeckungsmaterial, gemessen an der Losungsablage. Pedobiologia 9: 120-127.

McIntyre, D. S. 1958. Permeability measurements of soil crusts formed by raindrop input. Soil Sci. 85: 185-189.

Scharpenseel, H. W. and H. Gewehr. 1960. Studien zur Wasserbewegung im Boden mit Tritium-Wasser. Z. Pflanzenernaehr. Dueng. Bodenkd. 88: 35-49.

Schwerdtle, F. 1969. Untersuchungen zur Populationsdichte von Regenwuermern bei herkoemmlicher Bodenbearbeitung und bei "Direktsaat". Z. Pflanzenkrankheiten und Pflanzenschutz 76: 635-641.

Teotia, S. P., F. L. Duley, and T. M. McCalla. 1950. Effect of stubble mulching on number and activity of earthworms. University of Nebraska, College of Agriculture, Res. Bull. 165, Lincoln, Nebr.

Wilcke, D. E. 1962. Untersuchungen ueber die Einwirkung von Stallmist und Mineralduengung auf den Besatz und die Leistungen der Regenwuermer im Ackerboden. Monogr. Angew. Entomol. 18: 121-167.

Williams, R. E. and D. W. Allman, 1969. Factors affecting infiltration and recharge in a loess covered basin. J. Hydrol. 8: 265-281.

B. FERTILIZERS

The use of organic fertilizers to increase earthworm populations has been strongly recommended by Dr. Barrett[4] and other organic writers, and the value of the practice has been confirmed by several scientific studies as well.[5] However, the effects of inorganic or chemical fertilizers on earthworm populations have been a matter of some controversy between organic and conventional farming advocates. While this controversy is not yet entirely settled, it must be said that the weight of evidence is tending to favor those who feel that certain chemical fertilizers will benefit, rather than harm, earthworms.[6] Such beneficial effects can be obtained from those fertilizers which tend to make soil more neutral, or less acid; and from Nitrogenous fertilizers which increase plant growth, thereby increasing earthworm food supplies. Fertilizers which increase soil acidity, on the other hand, such as sulphate of ammonia, can be quite harmful.

We have already noted in chapter 2 that earthworms can play a major role in incorporating organic nutrients into the soil. Such nutrients can be naturally occurring leaf litter or plant residues, or manures added by human hands. We have also noted that earthworms transform this material into new forms more useable by plants. This latter opinion was confirmed by Dr. Maria Poloozsny in an article published in *Pedobiologia* in 1975. In the English summary of that article, she said,: "The result of the chemical analysis of the experimental soil and (castings) proved that not only a mixture of the raw organic material with the mineral soil is accomplished, but also a change of quality of the organic matter. From this practically humus-free loess (windblown soil), develops from (castings) materials characteristic of natural soil."[7]

C. IRRIGATION

One practice which can clearly be beneficial to earthworms is the use of irrigation to supply additional moisture to soil where insufficient amounts of water are provided by natural rainfall. Since it is well established that earthworms hold up to 85% of their live weight as water, and that moist environments are essential to earthworm survival, the benefits of irrigation to earthworm populations should not be surprising. However, it should also be noted that the presence of earthworms can also be beneficial to the farmer doing the irrigation.

It should follow from what we know about increases in soil water holding capacities and infiltration rates that irrigation water applied to earthworm populated soils will be used more efficiently and completely than the same volume of water applied to soils without earthworms. And indeed, Frank Hinckley, the orange grower whose experience with earthworms is recounted by Dr. Barrett,[8] reports, "In regard to the amount of water used, I find that since the worms have opened up the soil water penetrates more freely. I can irrigate in a much shorter time and with a larger volume of water per furrow. Under this method, of non-cultivation, I use a little less water, but the trees are able to use more of that which is applied." As Mr. Hinckley's experience has been confirmed by "hard scientific" research, it is difficult to overstate the importance of such a finding to semi-desert and other arid areas where water is in short supply and expensive to obtain and apply.

It has been observed by researchers in Australia that several changes in earthworm populations and activities are associated with opening up lands to new irrigation programs; including an increase in biomass (organic matter), and shifts among population densities of different earthworm species, as well as a gradual spread into new areas of habitation, particularly along irrigation bays. It has also been noted that earthworms, after they become established, gradually diminish the density of mats of unincorporated organic matter, by working them into the soil. The net effect of this mat incorporation on pasture yield in Australia's Riverine Plain has not yet been established.[9] But where soil mats continue to increase, fewer and fewer seeds germinate or young plants manage to struggle through to mature.

D. PESTICIDES AND HERBICIDES

Perhaps no subject related to farming and gardening causes more heated argument between "organic" and "conventional" farming advocates than the use or nonuse of chemical biocides to control harmful plants or insects. The settlement of such a controversy goes far beyond the scope of this book, and we shall not comment on the general aspect of this matter.

There is a fair amount of scientific literature available on the effects of various pesticides on earthworms, and it shows that these effects are quite variable, from zero mortality to 100% mortality, depending on the pesticide substance studied. Therefore, it is not possible to make blanket statements about the effects of pesticides in general on earthworm populations. It must also be noted that pesticide effects will vary not only with the substance applied, but also with the method and rate of application, the time of year, temperature, and a number of other elements.

The table below is compiled from a large number of studies, and provides an overview of different pesticides in terms of their effects on earthworms.

It may be worth noting that more than half of the

TABLE 7.6—Thirty-four pesticides and their toxic effects on earthworms

Substance[10, 11, 12]	Toxic Effect	Reference
AC92100	Heavy	10
Aldicarb	Heavy	11
Aldrin	Slight	11
Azinphosmethyl	None	11
BAY 37289	Slight	12
BHC	Slight-Moderate	11
Bux	Moderate	12
Carbaryl	Heavy	11
Carbofuran	Moderate-Heavy	10-12
Chlordane	Heavy	11
Chlorfenvinphos	Slight	11
Dasanit	Heavy	12
DDT	None-slight	11
Diazinon	None	11
Dieldrin	Slight	11
Disulfoton	Slight	11
Dursban	Slight	11
Dyfonate	Slight	11
Endrin	None-Moderate	11
Fenitrothion	Slight	11
Guthion	None	11
Heptachlor	Heavy	11
Leptophos	Slight	10
Malathion	None-Slight	11
Menazon	None	11
Methomyl	Moderate	10
N-2596	Heavy	10
Parathion	Slight-Moderate	11
Phorate	Heavy	10-11
Sumithion	Slight	11
Telodrin	None-slight	11
Trichlorphon	Slight	11
WL-24073	Moderate	10
Zinophos	Moderate	11

substances shown in the above table have no or only slight toxic effects on earthworms. Therefore, it is apparent that a farmer or gardener does not necessarily have to abandon chemical aids in order to obtain the benefits of adding earthworms to his fields or plots; although he must be selective in choosing which chemicals to use, and must use them in some moderation. It would seem that both organic and chemical methods need to be approached on a case-by-case basis, and not on the basis of wholesale prejudices.

Several studies have shown that earthworms have a tendency to concentrate certain chemicals in their tissues, including DDT and other toxic materials. Such concentration may pose a problem where earthworm populations are very heavy, and where earthworms form a significant link in the food chain of birds or other animals which may lead up to human beings. Dr. Charles Gish, a scientist working for the U.S. Dept. of the Interior, conducted a study in 1965, published in 1970[13], which suggested that, since the ratios of certain chemicals in earthworm tissue to amounts found in the soil from which the earthworms were taken were relatively constant, earthworms might be useful as monitors of pollution levels affecting rural land areas. Further studies exploring this concept have produced very promising results at the University of Wisconsin Dept. of Soil Science, and also at Xavier University in New Orleans.

E. EARTHWORM MIGRATIONS

Taxonomists and other soil scientists now postulate that earthworms and human beings have been affecting each other since Man first engaged in the growing of crops and the rearing of domestic animals. And one of the effects that Man has had on earthworms over the centuries is to spread certain dominant species from their points of origin to many distant lands.

Most of the dominant worms in the U.S. today are not native to our continent, but rather are "exotics" imported from Europe, probably accidentally by very early colonists. The same process seems to have occurred wherever European colonists settled, from Maine to Australia. Earthworms probably traveled in a variety of conveyances, from sheltered planting pots to the hooves of horses. G. E. Gates did one study in which he found certain species of worms associated with greenhouse activity, with some worms remaining in the greenhouses themselves, and others migrating outward into the surrounding soil.[14]

While earthworms act to reduce erosion in some soils, in other situations, erosion of the soil takes the worms along with it. Dr. Atlavinyte reported that 36.2% of the earthworms under fallow soil, 16.9% of those under arable land, 7.6% of those under perennial plants, and 4.5% of those under sod were carried off by erosion in a single year.[15]

It is clear than Man affects the Worm every bit as much as the Worm affects Man. The question before us now is whether we can learn to cooperate with this life form so that the Worm can affect us more positively; and we can do the same for the Worm. Much depends upon your reply.

★ ★ ★

REFERENCES

1. Edwards & Lofty, **BIOLOGY OF EARTHWORMS**, op. cit. pg. 174-5.
2. Ibid, pg. 176.
3. Ehlers, W. & Baeumer, K. 1974. Soil moisture regime of loessial soils in western Germany as affected by zero tillage methods. *Pakistan J. Sci. Ind. Res.* 17 (1):·32-29.
4. Barrett, Thomas J., **HARNESSING THE EARTHWORM**, op. cit. pg. 45.
5. Edwards & Lofty, **BIOLOGY OF EARTHWORMS**, op. cit. pg. 133-136.
6. Ibid. pg. 178-80.
7. Poloozsny, M. 1975. Die bedeutung zwei regenwurm-arten fur humifizierungsprozessel. *Pedobiologia* 15: 439-45.
8. Barrett, Thomas J. **HARNESSING THE EARTHWORM**, op. cit. pg. 59.
9. A) Barley, K. P. & Kleinig, C. R. 1964. The occupation of newly irrigated lands by earthworms. *Aus. J. Sci.* 26: 290-1. B) Rixon A. J. 1968-9. Irrigated riverine soils are changing. *CSIRO Rural Research.* (Commonwealth Scientific and Industrial Research Organization, Australia). C) Noble, J. C. & Mills, P. A., 1974. Soil moisture status and its effect on earthworm populations under irrigated pastures in southern Australia. *Proc. 10th Intl. Cong. Soil Sci.* Moscow.
10. Tomlin, A. D. & Gore, F. L. 1974. Effects of six insecticides and a fungicide on the numbers and biomass of earthworms in pasture. *Bull. Environmental Contamination & Toxicology* 12 (4): 487ff.
11. Edwards & Lofty, **BIOLOGY OF EARTHWORMS**, op. cit. pg. 182-4.
12. Thompson, A. R. 1971. Effects of nine insectides on the numbers and biomass of earthworms in pasture. Bull. Env. Cont. & Toxicol. 5 (6): 577-586.
13. Gish, C. D. 1970. Organochlorine insecticide residues in soils and soil invertebrates from agricultural lands. Pesticides Monitoring J. 3 (4): 241-252.
14. Gates, G. E. 1963. Miscellanea megadrilogica VII. Greenhouse earthworms. Proc. Biol. Soc. Wash. 76: 9-18.
14. also see Gates, G. E., 1976. More on oligochaete distribution in North America. *Megadrilogica* 2 (11): 1-6.
15. Atlavinyte, O. & Kuginyte, Z. & Pileckis, S. 1974. *Pedobiologia* 14: 35-40.

CHAPTER 8

ALL ABOUT CASTINGS

A. HISTORY

The Rev. Gilbert White wrote in 1775:

> "Worms seem to be the great promoters of vegetation, which would proceed but lamely without them, by boring, perforating, and loosening the soil, and rendering it pervious to rains and the fibres of plants, by drawing staws and stalks of leaves and twigs into it; and, most of all, by throwing up such infinite numbers of lumps of earth called *wormcasts,* which, being their excrement, is a fine manure for grain and grass."[1]

So acute were Dr. White's powers of observation that nearly 200 years later, we still know little more about earthworm castings than the essential information he provided then. The testimony as to the effectiveness of earthworm castings as planting media and as soil amendment from organic gardeners and farmers is a chorus of overwhelming volume. These gardeners report that earthworm castings improve lawns, corn, begonias, violets, rubber plants, watermelons, pumpkins, roses, etc. The following letter, written in 1975 by the Garden Editor of the *Las Vegas Review Journal,* is typical.

Dear Mr. Gaddie: Nov. 15, 1975

    This letter is really a testimonial for the use of earthworm castings to produce super flowers and vegetables.

    In March of this year I was able to obtain a truck and trailer load of earthworm castings for a rose garden our women's club was getting ready to build and present to the City of Las Vegas in Lorenzi Park, a city-owned park. Mixed with the castings were several bales of peatmoss, sacks of steer manure and enough sulphur to neutralize the alkali present in our Southern Nevada soil and water.

    We were holding 204 bare root rose bushes and time was getting very late when the load of castings arrived. The four beds being planted totalled 80 feet long and 40 feet wide and the soil underlying all of it consisted of pure clay. The beds were bordered with 10 inches of cement curbing. 10 inches of the clay was taken out of all the beds and then separate holes dug to a depth of 14 inches for each rose.

The beds were filled to within 4 inches from the top of the curbing as the roses were planted in each hole and the holes also filled with this rich material.

Ordinarily there is almost a 50% loss of bare root material in our area when it is planted that late, but in this case by August we had suffered only a 5% loss. Also usually our roses go dormant, more or less, during July and August due to extreme heat, but in this case the roses bloomed and grew without stopping clear through. So many roses were produced that we had to prune every two weeks. The flowers are just as vigorous at this date and yielding pounds of petals for drying and the making of potpourri as presents when the rose garden is dedicated in April of next year.

In July we put a thin mulch of sawdust all over the beds and a light feeding of Bandini was applied, although no sign of lack of food was evident. It was applied on the advice of Armstrong's where the roses were purchased. The advice was based upon the usual soil and not where such large amounts of earthworm castings were used. From the appearance of the roses it looks as if they wouldn't have had to be fed until this coming spring.

The beds have been flooded a week and this, with the rich soil, has produced this beautiful garden.

PHOTOGRAPH 8.1 – Lorenzi Park Roses
(Photo Courtesy of Adelene Bartlett)

No black spot, mildew or aphids have invaded any of the beds and many "basal breaks" have occurred, which shows the bushes are healthy and bursting with life.

We cannot praise earthworm castings highly enough and hope that much more of it will be available in the coming years to the general public and that the value of it will become more widely known.

Sincerely yours,

Adelene Bartlett

Adelene Bartlett
Co-Chairman Mesquite Club
Lorenzi Park Rose Garden
1825 Bracken Ave.
Las Vegas, Nev. 89104

Despite the abundance of testimony such as the letter above, little scientific research has been done on the question of the effectiveness of castings as planting media or soil amendments. In fact, a study initiated in 1976 by North American Bait Farms and a large nursery, comparing earthworm castings to conventional planting media, is thought to be the first work of its kind undertaken so far.

B. RESEARCH

While there is a lack of direct scientific evidence, a strong case for the usefulness of castings can be made on inferential grounds. In the first place, the effectiveness of earthworms as promoters of vegetation has already been established, and we know that a considerable portion of their activities benefit plants through their castings. It therefore seems reasonable that the castings alone would also be beneficial to plants.

Secondly, the known characteristics and composition of earthworm castings are similar to the characteristics and composition of planting media and soils of proven fertility. Lunt & Jacobson produced the classic study on earthworm castings in 1944, entitled: *The Chemical Composition of Earthworm Casts.*[2] The findings of this study are well known and will not be repeated here. The study itself is reprinted as part of Thomas Barrett's book, **HARNESSING THE EARTHWORM**, p. 128.

Edwards and Lofty summed up the characteristics of castings in these words: "Earthworm casts contain more micro-organisms, inorganic minerals and organic matter in a form available to plants, than soil. Casts also contain enzymes such as proteases, amylases, lipase, cellulase and chitinase, which continue to disintegrate organic matter even after they have been excreted."[3]

It should be noted that different lots or batches of castings may vary in their contents, depending on the species of worm producing them, feeding material (local soil composition) and other variables, age since excreted, etc. In general, castings samples have been found to contain 0.5-2.0% Nitrogen; .06-.68% phosphate phosphorus; .10-.68% potassium; .58-3.50% calcium; and other nutrients in varying proportions.[4] Because earthworm castings generally total less than 5% NPK, they cannot be sold in some states as organic fertilizers; but must be marketed instead as a soil amendment or agricultural mineral.

C. HOW TO PRODUCE EARTHWORM CASTINGS

The following information is provided for the commercial earthworm grower who is interested in increasing his production of earthworm castings. We will assume that such a grower has already obtained a copy of VOL. I of this series, *Scientific Earthworm Farming*, and that he has been conducting his operations according to the recommendations contained there.

Assuming you have an established bed containing at least 100,000 worms (4,000 worms per cubic foot), and that your bed has been operating for at least 90 days, you can produce a large quantity of castings in only a few weeks by following the steps below:

1. Feed heavily for one week; 50-60 lbs. dry weight of cow manure, which should weigh about 180 lbs. wet.

2. During the second week, turn the top 6 inches of bedding but do not feed any additional manure or supplement.

3. Turn the entire bed. At the start of the 3rd week, spread 3 lbs. of a high grade protein supplement such as WORM-GRO on the bed. Repeat this process with an additional 3 lbs. 3-4 days later. This supplement will help to increase the nitrogen content of your castings. DO NOT FEED MANURE DURING THIS WEEK. Use only the supplement.

4. Three to 4 days after the last supplement feeding, remove the top 3 to 5" of bedding, which will contain virtually all of your worms, and place them in another bed. The remaining 6-9" will be earthworm castings. Remove the castings and screen them with a 1/8" screen to remove any undigested material. Sun dry the castings.

A standard bed will consume an average of 2,080 lbs. of manure dry weight per year, which will equal 7,956 lbs. wet weight. This material will yield about 1,200 lbs. of dried castings per year.

D. MARKETING CASTINGS

Castings can be marketed in a variety of ways depending on the size of your operation, the amount of money you wish to invest in packaging and advertising, shipping facilities, etc. Castings are generally sold by volume today rather than by weight, since moisture contents may vary and your weight will change with the amount of moisture. You can market castings at retail in 1 quart, 2 quart, 1 gallon, or 2 gallon (1 peck) packages. Or you may sell castings wholesale by the cubic yard. One cubic yard of castings will weigh between 1,000 and 1,300 lbs., on the average.

The price you receive for your castings will depend on several factors including:

1. Whether you are selling wholesale or retail.
2. The price of other planting mixes, and soil amendments available in your area.
3. The volume of castings being purchased.
4. Whether you are selling pure castings or a mix of castings and other material.

Many wormgrowers mix their castings with dried poultry manure, peat moss, sand, vermiculite, or other materials; depending on the use to which the castings will be put, quantities available, and other factors. Castings which will be used in a planting mix may be joined with peat moss or other lightweight substances to reduce the weight/volume ratio. If they are to be used as a soil amendment, they may be mixed with dried poultry waste or other high protein organic materials to increase the NPK value of the material.

You should, as a rule, avoid trying to sell castings as a fertilizer per se, since many states have strict labeling and other requirements for materials labeled as fertilizers. You should check with your State Department of Food and Agriculture, or your local Cooperative Extension Service office (County Agent) regarding regulations in your area.

If you are only producing a *small amount* of castings, you may be well advised to limit your marketing program to friends and neighbors, since the demand for castings has been known to skyrocket rapidly when they are sold through established retail channels and you could find yourself with more demand than you can supply.

If you have a *moderate size* operation, say 100-200 beds, you could develop a simple package and offer it to local independent retail nurseries. Recognize that you will have to compete against other planting media in their stores, and your product should be priced at not more than 120% of the average price for these other products. From this price (retail), you will discount between 40-50% to the nursery. For example, if you feel that your castings should sell for $1.00 per quart retail, you would receive between 50 cents and 60 cents for the same package at wholesale. Naturally, you must then calculate your costs of production, packaging, and delivery to assure yourself that you will be making a reasonable profit.

If you have a *very large* operation, 500 beds or more, or if you are part of a marketing cooperative with other worm-growers, you may want to consider offering your product to a large wholesale nursery grower, or a local chain of retail outlets. In this case, you should approach the nursery with an offer to give him a reasonable quantity of castings for a test planting program. Again, it will be important to know what other materials the nursery is now using, and what the cost of those materials is. The primary interest of most wholesale nurseries in using castings will be to reduce their costs of operation, rather than to obtain faster or healthier growth of their plants. Again, you should find out how much planting material the nursery will be using over the course of a year, and compare this to your own production capacity. One nursery in Southern California is now using 75,000 cubic yards of planting mix per year. If they were to substitute castings for their present mix, they would consume the entire castings production of nearly 100,000 earthworm beds!

Because only the very largest operations are equipped to provide the huge quantities of castings required by national firms, most growers should probably not waste their time approaching such accounts. One major fertilizer company reports, that they receive an average of 3-6 calls per week, from growers with less than 100 bins wanting to supply them with castings. As a result, they almost gave up hope of finding a supplier actually able to meet their needs.

Of course, if you are affiliated with a reliable contract wholesaler, you may find that the wholesaler is already developing plans for marketing your castings. In such a case, your marketing problems are already solved.

E. USING CASTINGS IN YOUR OWN GARDEN

In general, castings can be applied to your lawn or garden the same way that other manures are applied: by spreading on top of the ground. The difference is that castings will not burn plant roots, since the raw nitrogen content is much lower than for many kinds of manure. Thus, you can use castings with more confidence and in greater amounts.

The Difference Castings Can Make

PHOTOGRAPH 8.2—The smaller one was grown in soil without earthworms, while the larger one was grown in soil rich with castings.

For lawns, you can spread castings to a depth of 1-2" before planting a new lawn, or to a depth of ½" on an established lawn. The same depth can be used for your vegetable garden: 1-2" before planting, or ½" after planting. In either case, if your soil is particularly deficient in nitrogen or other nutrients, you may wish to add some cow, horse, or poultry manure with the castings. However, these materials can burn plant roots, and should be used with caution. Many gardeners, even with the worst of soils, report that they receive the best results from castings alone.

Indoor Plants

A favorite use of castings for indoor plant enthusiasts, is as a planting mix; either alone or with peat moss or vermiculite. Here you simply fill the pot or window box with castings to within 2" of its top. African violets are a favorite indoor plant, but one which is extremely delicate, requiring expensive commercial potting soils to survive. Yet one grower reports that he had hardier and more fully blooming violets with pure castings than with any mix he had ever used!

Castings are frequently used as "medicine" for sick or dying indoor plants. In this application, simply remove 1-2" of the old planting mix (being careful not to disturb the roots of the plant more than necessary), and replace with earthworm castings. Such a procedure has been known to effect many "cures" of the most sickly plants.

More Advantages

Castings can also be used to make up the planting root ball for newly planted fruit trees, rosebushes, berry vines, etc. This use of castings appears to provide a "head start" for the plants which continues to produce remarkable advantages even years later.

It should be noted that earthworm castings are more porous than most soils, and will therefore require more water initially. However, the castings will also hold more water than conventional soils, and therefore may need watering less frequently. You should check your plants carefully when using castings, to be sure that you are neither over nor underwatering.

Since castings are a material extremely rich in humus (the

final stage of organic matter decomposition), and since the mineral nutrients are in a slow release form in most cases, plants in castings will probably need less food or fertilizer than those in conventional soil or planting media.

This letter explains the Staples system of gardening which successfully uses worm casts.

San Bernardino, Ca.
March 12, 1976

Dear Mr. Gaddie:

We wish to share our experiences in using worm castings in home gardening by employing a method which has been successful under adverse soil conditions.

This system employs the easy gardening approach which has the following advantages:

1. No commercial fertilizers
2. No commercial or poisonous sprays
3. No digging
4. Very little weeding
5. Better production in a small area
6. Natural beauty
7. Recycling trash or garbage to natural soil

We apply approximately 6 to 8 inches of worm castings directly to the soil without any prior soil preparation. This application supplies enough nutrition to adequately supply the plants throughout the growing season if the worms are fed as suggested in your book "Earthworms for Ecology and Profit".

Seeds may be planted directly into the worm castings or young plants transplanted into worm castings without fear of damage to the plants that may be experienced by those using other types of fertilizers or manures. We have also noticed a minimum amount of transplant shock when transplanting flowers, vegetables, shrubs, herbs, or trees directly into very wet worm castings, without the use of commercial additives.

Since the worm castings cover the entire area to be planted, we are able to place our plants closer than normally recommended because we know that there is sufficient nutrition to sustain this intensive planting. We place our plants in small groups, not to exceed 3 feet in any direction, rather than in rows. Various types of plants, such as herbs, flowers, and vegetables, are alternated for insect control. It seems that insects find their favorite plant by their sense of smell. We confuse the insects by alternating vegetables with marigolds, garlic, mints, etc., which keeps our insect damage to a minimum without using commercial sprays.

When our plants have grown to a height of 4 to 6 inches, we add an organic mulch such as straw, cardboard, leaves, cured grass clippings, or newspaper. This mulch serves various purposes. First, weeds are discouraged thus reducing a disagreeable garden chore; water is retained by less evaporation of moisture from the worm castings; and the organic material may be used as food by worms which remain in worm castings.

This "method" may be used as a yard border, on a patio, in a window box, or in any space, however limited. By combining flowers and herbs with the vegetables, there is a natural beauty which will not be found in a more conventional garden.

Organic material which would normally be disposed of as trash or garbage may either be introduced into the worm beds to be reduced to worm castings or used as mulch on the garden. This procedure contributes to an ecology balance and saves money for the home gardener.

Sincerely,

Elgin L. Staples
Elgin L. Staples

W. Joan Staples
W. Joan Staples

F. THE FUTURE OF CASTINGS

The role that earthworm castings can play in commercial horticulture and agriculture in the future will depend on two factors: research and economics. Some organic farmers operating on a moderate scale are reported to be using castings in commercial production now, and to be receiving excellent results, particularly with vegetable crops such as corn, tomatoes, squash, and pumpkins. But until large scale planting trials can be completed at university based agricultural research centers, this practice is not likely to be widely adopted by farmers in general. We believe that the anecdotal evidence from the organic tradition regarding castings is now sufficient to justify the commencement of such trials, and vermiculturists should seek out opportunities to promote this kind of research with county agents and local garden clubs.

The question of economics is another matter, and one which may be more complex in its solution. If castings are to be used on a large scale in commercial farms, they must not only be proved effective in controlled planting trials; they must also be able to compete on a cost basis with chemical fertilizers and other forms of organic fertilizer, such as dairy manure and poultry waste. Although the cost of chemical fertilizers has been rising in recent years, as natural gas and other petroleum feedstocks become more expensive and less available, it may be some time before they will increase to a point where they are more expensive than castings, which are presently being produced only on a small scale. The production of castings in annelidic consumption facilities processing biodegradable residential refuse could change this picture, and this operation will be discussed in the following chapter.

Also, final note should be made on the problem of the too limited view of some farmers when considering various fertilizer sources. Many farmers, even today, are convinced that the only value a fertilizer or soil amendment has is its content of nitrogen, phosphorus, and potassium. But modern agronomists are aware of the immense importance of a range of other nutrient materials, the micro-nutrients which are increasingly scarce in many agricultural areas, but which castings have in good proportions. The effect of castings on soil pH, and soil structure and water holding capacity should also be considered. In short, castings need to be considered as part of a total plant growing environment, and when they are so considered, they will be regarded favorably by increasing numbers of farmers in the future.

★ ★ ★

REFERENCES

1. White, G. H., 1775. The Natural History of Selbourne, England, pg. 213-4.
2. Lunt, H. A. & Jacobson, H. G. M. 1944. The chemical composition of earthworm casts. *Soil Science* 58 (6): 367-375.
3. Edwards & Lofty, **BIOLOGY OF EARTHWORMS**, op. cit. pg. 121.
4. Figures supplied by North American Bait Farms, Ontario, Calif. for castings samples from their farm.

CHAPTER NINE

EARTHWORMS AND WASTE DISPOSAL

The problem of solid waste pollution, although less widely appreciated than the problems of air and water pollution, may be even more serious than either of these. We are currently producing 135,000,000 tons of solid waste every year in the United States alone. By 1985, this is expected to increase to over 200,000,000 tons per year, or nearly 1 ton for every man, woman, and child, now living in this country.

A. SANITARY LANDFILLS

The predominant method now being used to dispose of this huge quantity of material at present is sanitary landfilling. Put more simply, we are burying it. But sanitary landfill is becoming increasingly expensive, hazardous, and generally impractical. In the first place, we are running out of land near cities on which we may dispose of waste. Such land that is still available is increasingly expensive. In the second place, it has been found that sanitary landfills are extremely hazardous to the environment. Water passing through a landfill, from rainfall or other sources, becomes polluted, and may in turn pollute the underground water table, or even large rivers or streams running nearby. In some landfills, toxic methane gas which is highly explosive can build up creating a considerable hazard of fire or explosion. Landfills which have been converted to parks or other uses, including building sites, after they have been filled, have been known to settle, causing considerable damage to property and hazard to life.

As though these hazards and disadvantages were not sufficient to doom the concept of sanitary landfill, there is another overpowering consideration: we simply cannot afford to waste all of the resources now going into these landfills because they are too valuable; and in some cases irreplaceable.

The federal government, as well as all state and local governments, are now spending billions of dollars every year to operate waste disposal facilities, and to find new and better methods of recovering the value contained in the resources we now throw away. The problem may be divided into two parts: 1.) *recovery* of nonbiodegradable resources such as metal, glass, and plastics from the refuse stream; and 2.) *use* of the biodegradable refuse, such as table scraps, animal and crop wastes, lawn clippings, newspapers, cardboard boxes and packages, etc.

B. SEPARATION OF WASTES

Recovery of the nonbiodegradable portion of our refuse is both an engineering and a social problem. The most efficient method of recovering this waste would be to keep it separate from the biodegradable waste right at the start. This is called home or source separation. Some communities have passed regulations requiring deposits on bottles and cans, and mandating home separation of other solid refuse, before it is picked up. Unfortunately, these approaches do not appear to offer a complete solution in most cases; because not everyone is willing to cooperate in such measures. Therefore, ways must be found to separate the "hard fraction" (nonbiodegradable) material from the "soft fraction" (biodegradable.)

Fortunately, the technology for this kind of separation is advancing rapidly. Air classification systems, centrifugal separators, even light diffraction and heavy media separation systems have been developed and are undergoing testing in a number of communities. It is probably safe to say that the separation problem will be solved in the fairly near future. It is still cheaper for each of us to do the needed separation at home.

Techniques, for reuse of most metals and glass, are fairly well known; and markets for these materials have been operating for many years. In most cases, the metal or glass is simply melted down and reformed into new metal or glass, which can be used in new products. But the question of what to do with the biodegradable fraction of our refuse is not yet settled.

C. WASTE INTO ENERGY

One kind of scheme which has attracted much attention, and enormous amounts of public and private capital in recent years, is the conversion of this biodegradable refuse into energy. 1.) In one process, the refuse, after being separated, is burned in very high temperature furnaces. The heat is used to boil water for steam, and the steam is piped to a generator to create electricity. 2.) Another process, called *pyrolysis,* involves baking the material in an oxygen-free chamber, to produce a synthetic fuel which can be burned in the same way as gasoline. 3.) Still another process, called *anaerobic digestion,* uses certain kinds of bacteria to convert the waste into methane gas, which can be used as a substitute for natural gas.

While such waste to energy conversion processes may offer some advantages in certain situations, there are two major problems associated with them. 1.) Problem number one is expense. Most of these systems involve construction costs of as much as $150,000,000 for a single plant. This is an enormous expenditure, far beyond the capacity of most private enterprises, which means that most of these plants will be built with your tax dollars, even when operated for private profit. Even the smallest system will cost well over $10,000,000. Another aspect of this expense is that the technology for most of these plants is still being developed, as the City of Baltimore recently discovered to its sorrow. Finally, there is a larger question: may we not find in a few years as solar and other energy sources are developed that we no longer need to burn waste for fuel? $150,000,000 is a lot of money to spend for a plant that may be obsolete in only ten or fifteen years. 2.) The second problem with using all of our biodegradable waste for energy is the fact that we also need this material as fertilizer. We have already noted the increasing scarcity of fossil based feedstocks for chemical fertilizers, and their increasing cost. There is every prospect that these trends will continue, until we reach a point at which these feedstocks (oil and natural gas) become unavailable at any price. 3.) Other processes, such as nitrogen extraction from the atmosphere, impose other hazards. A scientist at Harvard University recently theorized that fertilizer manufacture by nitrogen extraction would soon remove as much nitrogen from the air as is removed by all natural processes in the world. He further theorized that these processes would cause major damage to the ozone layer in our atmosphere.

As we can see, the problem of solid waste disposal is both

difficult and complex. However, we believe that a major part of the solution to this problem is literally lying at our feet: the earthworm. The process of using earthworms to consume biodegradable solid refuse and turn this material into castings which can be used for soil improvement is called "annelidic consumption." Facilities utilizing this principle are already operating on a commercial scale in both Japan and in Canada, and experiments are already under way to bring this process to an operational state in our own country.

D. HISTORY

The basic process of annelidic consumption of plant litter was described in detail by Charles Darwin in 1881.[1] The operation of this process in nature has been further documented by many scientists since. Dr. Thomas Barrett described how this process could be used on a farm in his book, **HARNESSING THE EARTHWORM**.[2] In 1954, Home, Farm, & Garden Research Associates published a book entitled, **LET AN EARTHWORM BE YOUR GARBAGE-MAN**, which described how this process could be applied in small scale home waste disposal systems.[3]

Ontario, Calif. (Cir. D. 28,786)
Daily Report **Experiment Leaves no Doubt** May 10, 1976
Earthworms Eat up Refuse

An Ontario city official says a four-month experiment leaves no doubt that *earthworms* can be counted on to consume common household refuse.

That conclusion could be significant since local officials are looking for ways to cut a possible major hike in cost of refuse disposal once the present disposal site fills.

In fact, said Assistant City Manager Dennis Wilkins, the success of the initial experiment has led to consideration of the possibility of another one on a larger scale.

Thousands of earthworms at North American Bait Farms, 1207 S. Palmetto Ave., Ontario, consumed most of 20,260 pounds of mixed household refuse delivered by the city last fall.

Frank Carmody, North American Bait Farms' director of public information, said 131 man-hours were needed to remove 1,858 pounds of metal, glass, plastics and rubber.

The remainder, 18,204 pounds of biodegradable refuse, was consumed by earthworms, he said.

Consumption of the rubbish, ranging from phone books to grass clippings was 50 per cent complete after 38 days and 80 per cent complete after 68 days.

The experiment concluded with a final inspection Jan. 28 by J. D. "Joe" Sallee, city refuse superintendent.

Wilkins said many uncertainties remain regarding how long consumption takes, what refuse processing is needed in advance and whether the whole idea is economically feasible.

Carmody raised other questions: How much land and equipment are needed? What is the most efficient flow of materials?

That such an operation would be sizable is shown by Carmody's estimate last fall that 300 million earthworms weighing 83 tons would be needed to consume Ontario's daily refuse output.

Even so, refuse is being successfully fed to earthworms by four commercial firms in Japan and there has been a small-scale experiment in Ontario, Canada, he said.

He revealed that a North American Bait Farms' proposal for a larger-scale experiment has been submitted to City Manager Roger Hughbanks.

He was reluctant to reveal details until Hughbanks has had time to review the proposal.

He said the firm would propose to share revenues from the experiment with the city and to set aside its own share for establishment of a full-scale processing facility.

Local officials have also been considering another possibility—a resource recovery plant that produces energy from refuse.

Wilkins has said such a plant might be economically feasible in the future if the cost of oil rises to the point that refuse is seen by Southern California Edison as an alternative.

Last year county supervisors scrubbed plans for a refuse dump in the Chino hills at the urging of Supervisor Robert Townsend.

He favored continued use of the Milliken Avenue landfill in Ontario, saying it could be used well into the 1980's when resource recovery may be feasible.

The first commercial scale annelidic consumption facility was established in Canada in 1970, and is currently processing about 75 tons per week of biodegradable refuse. Similar facilities were established in Japan in 1974 and 1975. At last report there were four such facilities, each processing about 10 tons per day. The facilities in Japan are primarily used for processing specialized manufacturing wastes, although a proposal for a facility to process residential refuse is now under consideration by several municipal governments in Japan.

In our own country, the first experiment designed to demonstrate the feasibility of the annelidic consumption process for residential refuse was undertaken in September, 1975, by North American Bait Farms in cooperation with the City of Ontario, California.

E. ONTARIO EXPERIMENT

The City of Ontario, California, like many suburban communities of its size (65,000 population), faces a serious problem in the future disposal of its household solid waste. Present disposal is by sanitary landfill on a site known as the Milliken Dump, operated by the County of San Bernardino, and located approximately 5 miles east of the city center. The Milliken Dump is approaching its maximum capacity and is scheduled for closing in 1981 or sooner. Because of the continuing population growth and development of the area, additional expansion of the Milliken Dump or location of an alternative sanitary landfill facility within a reasonable distance of the city center are both considered impractical. Capital and operating costs for waste to energy conversion facilities proposed to date would require substantial increases in household refuse disposal charges to Ontario residents.

Contact was made in August, 1975, between North American Bait Farms research staff and the Refuse Superintendent of the City of Ontario. After preliminary discussions with the Superintendent and other elected and staff officials of the City, an agreement was reached for the conduct of a joint experiment in annelidic consumption by NABF and the City of Ontario.

The Method

On September 22, 1975, the City of Ontario delivered 20,260 lbs. of mixed household refuse to the NABF experimental site. The refuse was hand sorted to remove all items of metal, glass, plastic, rubber, and other nonbiodegradable materials. Sorters removed 1,858 lbs. of such items, leaving 18,402 lbs. of biodegradable refuse for the worms to consume.

After separation, the biodegradable material was laid down between previously established beds containing earthworms and horse manure. These beds have been established for approximately 24 months prior to the commencement of this experiment, and had been fed cow manure until 1 month prior to the beginning of the experiment. The worms were starved for the last month before the refuse was laid down.

As the refuse was laid down in between the rows, it was thoroughly watered, covered with a layer of straw, and then covered again with wooden pallets to prevent movement of straw or refuse. It was noted that both odor and fly problems vanished as soon as the covering process took place, and subsequent inspections by the San Bernardino County Department of Public Health revealed no recurrence of these problems.

Once established in the rows, the refuse went through an initial period of bacterial decomposition lasting approximately 9 days. During this period, temperatures as high as 140 degrees F were recorded. However, these temperatures were confined to the refuse rows, and had no discernible effects on the earthworms in the previously established beds. Because of this heating factor, the earthworms remained in their beds during this period.

On October 1, 1975, temperatures in the refuse rows were recorded at 84 degrees F, and initial movement of earthworms from the adjacent beds into the refuse rows was observed. During the months of October, November, December, and January, 1976, periodic measurements and observations were made of the progress of decomposition of the refuse by the earthworms. By January 28, refuse consumption was estimated to be more than 95% complete, for virtually all of the biodegradable material, except for solid wood wastes, such as branches of trees.

The experiment showed that earthworms could consume biodegradable residential refuse in multi-ton lots, with feasible land, water, and energy requirements. While NABF has declined to release all of the measurements involved in the experiment for competitive reasons, it has been calculated that a 200 ton per day facility would require less than 100 acres of land to operate (land which would never fill up as in conventional landfill operations), and that such a

PHOTOGRAPH 9.2–Gradually all the litter was reduced to fine castings.

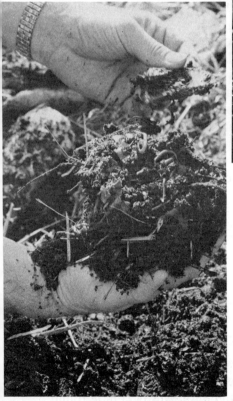
PHOTOGRAPH 9.1–The earthworms slowly munched their way through the trash.

facility could be established for less than $5,000,000 including land, equipment, and starting stock of earthworms. Since the earthworms do virtually all of the work of continually turning and aerating the refuse, the only energy consumed is in laying down the refuse originally and picking up the castings, and in providing perimeter lighting for the wormrows.

Annelidic consumption facilities offer a particularly attractive technology to smaller suburban and rural communities, generating refuse in lots of less than 1,000 tons per day; which is the minimum input required for feasible operation of waste to energy conversion facilities. A facility handling 20 tons per day by annelidic consumption could probably be established on less than 20 acres for less than $1,000,000 in startup costs. Operating costs for these facilities have been estimated as being comparable to costs for conventional sanitary landfill facilities now in operation; with the additional advantage that earthworm castings derived from the process may be salable to nurseries or farmers, providing revenues which can offset a considerable portion of these operating costs.

North American Bait Farms is continuing its work with the City of Ontario, and has since been contacted by no less than 8 other local waste disposal authorities, and at least one foreign government. NABF is willing and anxious to work with interested local authorities anywhere in the United States. You can make a major contribution to the ecology of your area by encouraging your local director of public works to write to:

> Director, Annelidic Consumption Programs
> North American Bait Farms
> 1207 S. Palmetto
> Ontario, CA 91761

F. ANNELIDIC CONSUMPTION ON THE FARM

Just as cities have struggled with increasing urgency in recent years to solve their solid waste disposal problems, farmers, especially those involved with livestock, have found themselves faced with increasingly serious problems of their own. The Environmental Protection Agency of the Federal

Government has grown increasingly concerned about the discharge of certain farm wastes, particularly dairy and feedlot manures, into water tables and streams. Here again, earthworms can help, as shown in this article from the Journal of Dairy Science, reprinted by permission of the American Dairy Science Association:

JOURNAL OF DAIRY SCIENCE
Vol. 55, No. 6

Biodegradation of Animal Waste by *Lumbricus terrestris*

Abstract

The possibility of recycling animal waste through the common earthworm, *Lumbricus terrestris*, was studied. Earthworms were raised in beds and fed only raw feces and water with lime added as a buffer. The conversion of kilograms of fecal dry matter to kilograms of live earthworms was 2:1. The excretion (castings) of the earthworm was a loose, friable humus type of soil containing 3.0% nitrogen. Earthworm meal dry matter analyzed 58% protein and 2.8% fat and proved to be very palatable when fed to domestic cats. Worm dirt was equal to greenhouse potting soil for the production of flowering plants. An added advantage is that the worm dirt weighs only about 50% as much as normal potting soil.

Introduction

It is not uncommon to find 1,000 dairy cows confined on a concrete lot of a few hectares. With an average weight of 635 kg/cow, a herd of this magnitude will defecate over 31.7 metric tons of raw feces daily. The disposal of the fecal matter without adding to the pollution of the surrounding soil, water, or air is a major problem that has not been solved.

Biodegradation of fecal matter by invertebrates is one of the proposed methods for coping with waste problems. Calvert et al. (2) used larvae of the common house fly, *Musca domestica*, to biodegrade fecal waste from caged laying hens. This process deodorizes and partially dries the feces in 2 or 3 days, and the resulting fly larvae can be used in animal protein supplements. Fresh feces containing 5.6% N are reduced to 2.0% N within 8 days by the larvae.

Biodegradation of feces by earthworms, *Lumbricus terrestris*, is another possibility. Earthworms can do an almost unbelievable amount of work. Darwin[3] in 1904 concluded that all of the vegetable mould in the country has passed through the intestinal canals of earthworms many times. Barrett[1] reported that as much as 269.1 metric tons of earthworm castings per hectare have been measured in a 6-month period in the Nile Valley.

Therefore, this experiment was designed to evaluate the ability of earthworms to convert organic waste materials into useful products.

Experimental Procedure

Twenty worm beds 3.66 m X 1.83 m X 0.457 m were established in an area that was shaded from direct sunlight by trees. The beds had temporary bottoms fashioned from galvanized iron. Cotton molt (fragments of seed, leaf, stalk, and lint removed from cotton by suction in the cleaning process) was placed in the bed to form a base about 5.0 cm to 7.5 cm deep. Worms from an established bed along with some of their original bedding material were then added and 1 week was allowed for the earthworms to get established before any fecal matter was added. Fresh feces were then scraped from the barnlot and spread over the beds in a layer approximately 7.5 cm thick. Worms were allowed to feed on the feces for a year with fresh feces being added in thin layers as needed. The beds were watered by natural rainfall and dry lime was added to keep the pH at 7.0. To control populations, the beds were first thinned by handpicking of adult worms at 45 days post-establishment and then monthly by clean picking for the remainder of the test. Random samples of feces, earthworms, and earthworm dirt were taken and dried at approximately 60-day intervals for analyses.

After 6 months the beds that appeared to be the most highly decomposed were chosen to be used as greenhouse potting material. Each week for 12 weeks 100 flowering plants were potted in earthworm dirt along with the same number potted in a normal potting soil by a local greenhouse. The soil was tested for nitrogen at the time of potting, 3 weeks later, and 6 weeks later. The plants were compared for appearance, root development, and weight by a panel of four individuals.

Samples of earthworms were taken at two different times, dried and ground in a blender, and fed to cats as a measure of palatability. To observe the range of organic waste materials that earthworms will readily degrade, the worms in one bed were offered cloth, dead animals, paper and burlap as well as feces.

Results and Discussion

Based upon the number of worms obtained by monthly thinning, growth and reproduction slowed during winter months; however, earthworms survived the winter climate of north Georgia without shelter other than the surrounding trees. The earthworms in the 20 beds thrived and grew rapidly in size and number. Worm numbers had to be reduced monthly to keep the beds from becoming over-populated. According to Swindle[4], worms thrive best in a near neutral medium containing 9.0 to 15% protein. Samples of bovine feces averaged 2.36% N X 6.25 = 14.75% protein. Lime was added to maintain a pH of 7.0. The feces contained considerable amounts of P, Ca, K, and Mg (Table I). Further analyses revealed 18, 256, and 96 ppm of Cu, Mn, and Zn.

Preliminary tests on small beds indicated that worms could convert raw feces to live earthworms in a ratio of 10:1. Mature cows excrete approximately 31.74 kg of fresh feces daily. Based on this, the raw feces from 1,000 cows could be converted to 3,171 kg of live earthworms daily (31.74 kg raw feces X 1,000 cows/10 = 3,174 kg). Earthworm samples averaged 22.9% dry matter (DM) which contained 58.2% protein, 3.3% fiber, and 2.8% fat (Table I). Thus, the feces from 1,000 cow herd would yield 423 kg of animal protein as dried earthworms daily (3.174 kg X 22.9% X 58.2% = 423 kg protein DM) or 154 metric tons of animal protein on an annual basis. Dried earthworm meal was offered to domestic cats for a palatability test. They quickly consumed their portions and begged for more.

TABLE I—Average mineral composition of cattle feces, earthworms and earthworm dirt (DM basis).

Source	Percentage				
	N	P	K	Ca	Mg
Feces	2.36	0.72	0.73	1.43	0.55
Earthworms	9.31	0.90	0.88	0.54	0.19
Earthworm dirt	2.98	0.32	0.40	1.20	0.36

Earthworm dirt was more porous and friable and weighed only one half as much as normal potting soil mixture. This difference in weight is an extremely important factor as it lowers the cost of shipping plants from one location to another.

The average analysis of the earthworm dirt was 3.0% N, 0.32% P and 0.40% K. One hundred flowering plants were planted in the worm dirt each week for 12 weeks. The same number were planted in the normal soil mixture to serve as controls. Six pairs of each 100 plants were put on a common watering system. There was no difference in growth rate, but the plants in the worm dirt had heavier root systems. A panel of four men subjectively evaluated six pairs of each 100 plants which were on separate watering systems, for growth, rate, root systems and number of blooms per plant. The plants in worm dirt required more water, grew faster, and had larger root systems and more blooms. The nitrogen content dropped from 3.0% to 1.25% N during the first 3 weeks, then down to near 0% during the next three weeks. This indicated that the nitrogen was readily utilized by the plants and that it was available to the plants over 6 weeks.

The worms in one bed were offered two dead ducks, burlap, cloth, and paper as part of their diet, and the first duck was completely devoured in 4 days. The second duck was much larger and 14 days were required for the earthworms to devour it. The worms readily consumed the samples of burlap, cloth, and paper. They will apparently deodorize and ingest a wide range of biodegradable materials and may be the solution to the large producer with a dead animal disposal problem.

Acknowledgement

The authors, **O. T. Fosgate** and **M. R. Babb** (Dairy Science Department, University of Georgia, Athens, 30601) are grateful to:

A. C. Cagle, Canton, Ga., who furnished the materials and facilities for this study.

References

(1) Barrett, T. J. 1947. **HARNESSING THE EARTHWORM.** Bruce Humphries, Inc.
(2) Calvert, C. C., N. O. Morgan, R. D. Martin and H. L. Eby. 1970. Biodegradation of poultry manure from caged layers. Porc. Poultry Waste Management Seminar, University of Georgia, Athens.
(3) Darwin, C. 1904. **THE FORMATION OF VEGETABLE MOLD.** Wm. Clowes and Sons, Ltd.
(4) Swindle, H. S. 1965. Commercial fishworm production. University of Georgia, Ext. Fish Manag. Mimeo. 21.

This principle of annelidic consumption can also be applied to crop wastes when it is not possible to simply turn them into the fields. Processing wastes from canning plants or other agricultural processing facilities can also be handled in this way, provided these wastes are pre-treated or diluted to remove excess acidity or other characteristics which may be harmful to earthworms.

G. SEWAGE & INDUSTRIAL WASTES

While annelidic consumption of municipal sewage is a theoretical possibility, and the technique has been used in some rural systems on a very small scale, there are some considerable practical difficulties to be overcome. Many sewage effluents or sludges, are filled with a variety of heavy metals which are toxic to earthworms and to many other forms of life. In such instances, some method of removing these metals from the sludge before feeding it to the worms must be found.

Except for agricultural processing wastes as noted above, the problem of developing annelidic consumption for manufacturing wastes is one which will require a great deal of research and development to solve. The technique has been used successfully on a commercial scale by some wood processing plants in Japan, and we believe the method does deserve further study for this type of application. But to expect that earthworms can solve all of these problems immediately is probably too optimistic. Man can reduce waste pollution by creating less waste himself. Increasing pollution is one unhappy by-product of a consumer society.

H. WORMS AND WASTE AT YOUR HOUSE

You need not wait for your city or town to adopt annelidic consumption in order to receive the benefits of having earthworms. Go to work on your personal household refuse. With a backyard Ecology Box, shown here, and about 20,000 worms to start, you can create a continuously functioning system to consume virtually all of the biodegradable refuse produce by the average family of four.

Plans and instructions for building and operating an Ecology Box can be obtained from the publisher of this book:

>Bookworm Publishing Company
>P.O. Box 655
>Ontario, CA 91761.

With this Ecology Box or a similar system, you can be turning your table scraps, lawn clippings, and paper wastes into earthworm castings; thus contributing to your personal environment and that of your community as well. You and your earthworms can decrease waste pollution.

REFERENCES

1. Darwin, C. 1881. **DARWIN ON EARTHWORMS**, op. cit.
2. Barrett, T. J., **HARNESSING THE EARTHWORM**, op. cit.
3. Home, Farm & Garden Research Associates, 1954. **LET AN EARTHWORM BE YOUR GARBAGEMAN**. Available from Bookworm Publishing Co.

★ ★ ★

PHOTOGRAPH 9.3—By the pond at North American Bait Farms

CHAPTER TEN

EARTHWORMS AS FOOD

NORTH AMERICAN BAIT FARMS, INC.

NEWEST DELICACY FROM THE GARDEN: VER DE TERRE RECIPE CONTEST ANNOUNCED[1]

North American Bait Farms, Inc., of Ontario, California, today anounced a world-wide search for recipes which call for "Ver de Terre" as an ingredient. "Ver de Terre" is the French name for earthworms. North American Bait is offering a prize of five hundred dollars ($500) for the best recipe submitted to their offices at 1207 S. Palmetto, Ontario, CA 91761, before October 31, 1975.

Purpose of the unique contest is to stimulate public interest in earthworms and their important contributions to agriculture and our environment, as well as their potential as a source of food for human beings. In announcing the contest NABF President Ronald Gaddie commented, "These little creatures offer many benefits to man which we have only begun to recognize. They can turn waste into fertilizer for plants, they can feed fish and other animals, and properly prepared they make excellent eating for people." Gaddie, a well-known connoisseur of "ver de terre", smiled and said, "I must admit, though, they are something of an acquired taste."

Like escargot (snails), ver de terre must first be washed in cold water and then boiled to remove stray bits of soil and to kill any undesirable bacteria. Properly prepared, ver de terre (earthworms) are an entirely safe and nutritious addition to human diets. Because ver de terre are entirely edible, with no bone or gristle to throw away, they are an excellent and potentially economical source of animal protein. The subtle flavor of ver de terre lends itself well to all sorts of other ingredients and methods of preparation.

Ver de terre can be served as part of any course, from the first appetizer to dessert. Recipes are being sought for such dishes as: "Canapes Ver de Terre," "Consomme Ver de

Terre," "Caesar Salada Au Ver de Terre" (using dried and crumbled ver de terre instead of bacon bits), "Ver de Terre Bourguignon," "Ver de Terre au Fromage Suisse" (baked ver de terre mixed with a sauce of melted Swiss cheese), "Ragout Ver de Terre," "Ver de Terre Forestier," "Ver de Terre au Truffe" (a gourmet's delight: ver de terre prepared with truffles), and "Petit Gateau au Ver de Terre" (redworm cookies, an old American country favorite).

Recipes and serving suggestions should be typed or hand printed on any kind of paper. Only one prize will be awarded, but authors of recipes selected for inclusion in a forthcoming book will receive $25.00 for each recipe used. As in all contests of this type, all entries become the property of North American Bait Farms, Inc., and none will be returned. Recipes will be judged for eye and taste appeal, ease of preparation, and economy of ingredients by a panel of food experts drawn from throughout Southern California. Judging will begin in November, and the winner of the contest will be announced in late December.

North American Bait Farms is an international company and a world leader in worm technology. Founded in 1967, the company sells tens of millions of worms every year, as well as a wide range of other products. A recently organized publishing division, known as **BOOKWORM PUBLISHING COMPANY**, publishes or distributes a full line of books on vermiculture (worm-raising), ecology, and related subjects. **EARTHWORMS FOR ECOLOGY & PROFIT**, recently issued, is already being hailed by vermiculturists (wormgrowers) throughout the U.S. as the new standard work in this field.

The use of earthworms as human food is a subject which has an enormous human interest value, to judge from the reams of publicity generated by a recent contest. As Dr. Taylor points out, it also has a certain validity in our increasingly hungry world. Earthworms have been used as human food by Maoris in New Zealand, in Japan, and in South Africa.[2]

But while people eating worms is interesting to think about, it is not too likely that earthworms will become a regular part of your diet or mine sometime soon. However, as a source of certain amino acids not as readily available from other protein sources in feeds for poultry, pets, or livestock, earthworms may play a significant role, in the near future.

A. PROTEIN

Up to 72% of the dry weight of an earthworm is pure protein, although this figure may vary depending on the species of earthworm and other considerations. However, at current prices, earthworms are too expensive to be used for crude protein. What is exciting to some nutritionists, though, is the fact that earthworms may be a prime source of certain key amino acids, not found in some other protein sources.

For example, earthworm protein is 10.07% arginine.[4] This is twice the percentage of arginine found in peanut meal, and three times the percentage of arginine found in anchovy meal. Earthworm protein is also relatively high in triptophan, at 4.41%, four times the percentage found in blood meal, and seven times the percentage found in beef liver. Earthworm protein is also 2.25 times as rich in tyrosine as liver.

One nutritionist has estimated that the use of earthworms to supply higher amounts of these critical amino acids, could result in a reduction of up to 70% in the grain feeding costs for certain types of poultry. This is possible because every animal must be given a proper balance among the various amino acids which compose the protein in its body. If this balance is not maintained, the animal may suffer serious diseases; and will almost certainly require more food because of less efficient digestion processes.

B. WILDFOWL DIETS

It was, in fact, the matter of digestion efficiency, which first led researchers in West Virginia to consider the possibility that earthworms play a significant role in wildfowl diets. It has been known for many years that quail, for example, require far more grain when raised in captivity than they do when subsisting in the wild. How was it possible for a bird to be every bit as healthy and hardy in the wild on a diet only one-third the size of that consumed by the same species of bird in domesticated flocks? The answer came when it was discovered that earthworms are an element in the natural diet of quail, but had been missing from the diet of those being raised domestically. The amino acid balances induced by consuming earthworms, allowed the quail to make far better use of the smaller amount of grain they could forage in the wild, since the amino acids from the worms made the quails' digestive systems more efficient.

C. "UNK UNKS"

Research on this topic is still in progress, and no reports have yet been published in the scientific literature. Nor have studies yet been completed on the relative economic efficiency of earthworms vs. other sources of amino acids. However, earthworms may have a significant advantage over artificial sources of amino acids: an advantage I call the "unk unk" factor.

"Unk unks" are what scientists call "unknown unknowns", and they have destroyed more than one elegant and expensive experiment. There are certain things we know, more things that we know we do not know, and a very large, perhaps infinite, body of things we don't know, and which we fail to realize we are ignorant of.

For agronomists of the 19th century, the role of micronutrients in plant growth was an "unk unk". Hazards of pollution with which we are intimately familiar today were "unk unks" to the men who made the industrial revolution. Animal nutritionists constructing feeding programs from artificial ingredients may also find themselves up against "unk unks."

The surest guide against the hazards of "unk unks" when dealing with living plants and animals is to imitate Nature whenever possible, and modify the natural process only after we are confident that this modification will indeed represent an improvement. The fact that many species of birds rely on earthworms as part of their natural diets, including chickens and other poultry, should in itself be grounds for intensive study of the earthworm to discover what values this lowly creature supplies for these birds. Nature does not proceed entirely by accident. If earthworms are consistently eaten by poultry with access to them (and we know that they are), there is a reason why this is so; and it would be to our profit to discover what that reason might be.

D. EARTHWORMS AS CATTLE FEED

Earthworms have also been considered as food for cattle, as part of an integrated recycling program of dairy grain feeds and wastes. Experiments recently conducted by the University of California at Davis explored the feasibility of using dairy cow manure as a supplement to add additional bulk to their diet, and to save on the costs of grain used to feed the cows. The experiments were based on the fact that bovine (cow) stomachs are not very efficient, and that there is a fair amount of undigested food value in most dairy manures; which can range as high as 15% protein. Unfortunately for the dairymen, the experiments showed that manure recycling didn't work. The cows didn't like the manure, and therefore wouldn't eat enough of it to maintain their weight and milk production. It was also found that the manure was too rough and mixed with too much dirt.

However, some scientists have suggested processing the manure by annelidic consumption, then harvesting the earthworms, grinding them up, and using them as a protein supplement. Since the earthworms would concentrate the protein from the manure, they would be a better source of nutrition than the manure itself. And because they could be processed in a more sanitary fashion, earthworms could be made more palatable to the cows.

E. PET FOOD

Earthworms have also been suggested as food for dogs and cats. As reprinted in Chapter 9, scientists at the University of Georgia found that cats found earthworm meal a very tasty dish. Some research on the possibility of using earthworms as a protein source for dog and cat food has been conducted at a large University in California, but much work remains to be done before this is likely to become a widespread use of worms.

F. COOKING WITH WORMS

If you're interested in becoming an ecological gourmet, you might want to try some "ver de terre" (that's French for earthworms) yourself. You might also try these recipes on friends at a Halloween Party. Following are the six recipes selected as finalists in the 1975 national earthworm recipe contest sponsored by North American Bait Farms. The contest drew more than 200 entries from all parts of the country, and will probably be an annual event. Each entry was evaluated by a panel of judges drawn from the faculty of the Nutrition Department of the School of Agriculture at California State Polytechnic University at Pomona. The judges rated each recipe for economy of ingredients, ease of preparation, and potential eye and taste appeal.

The judges then selected 12 semi-finalist recipes, which were prepared and tasted by a second panel of judges. The six best were then chosen for the final round of judging which was held in December at the Kellogg West Conference Center in Pomona. Judges of this final round included Dr. Ronald Taylor, author of **BUTTERFLIES IN MY STOMACH**, Dean Sander, News Director of Radio Station KLAC in Los Angeles, Dr. Ruthe Davis, Assoc. Professor of Nutrition at Cal Poly, and myself, Ron Gaddie. Here are the recipes they tasted:

APPLESAUCE SURPRISE CAKE

½ cup butter
1½ cup sugar
3 eggs
2 cups sifted flour
1 tsp. baking soda
1 tsp. cinnamon

½ tsp. salt
½ tsp. nutmeg
½ tsp. cloves
1½ cups applesauce
1 cup chopped, dried earthworms
½ cup chopped nuts

To prepare earthworms, chop coarsely and spread on Teflon cookie sheet. Place in 200 °F. oven for 15 minutes, remove and let cool, Meanwhile, cream butter and sugar, add eggs. Sift together the dry ingredients and add to egg mixture. Add applesauce, earthworms, nuts and mix well. Pour into a well-greased 10 inch tube or Bundt pan. Bake at 325 degrees for 50 minutes.

Mrs. P. H. Howell, W. St. Paul, Minn.

★ ★ ★

EARTHWORM PATTIES SUPREME

1½ lbs. ground earthworms
½ cup melted butter
1 tsp. grated lemon rind
1½ t. salt
½ tsp. white pepper

1 egg, beaten
1 cup dry breadcrumbs
1 Tbs. butter
1 cup sour cream
2 Tbs. plain soda water
   (for lightness)

Combine earthworms, melted butter, lemon rind, salt and pepper. Stir in soda water. Shape quickly into patties. Dip patties into beaten egg and breadcrumbs. Heat butter and cook patties in it about 10 minutes, turning once. Transfer patties to hot serving dish. Stir sour cream into skillet and heat thoroughly, pour over patties. Serve with plain boiled potatoes.

Mrs. P. H. Howell, St. Paul, Minn.

★ ★ ★

EARTHWORM OMELETTE

6 eggs
3/4 to 1 cup fresh earthworms
1/3 cup milk
¼ cup parsley
¼ cup sliced celery
1/3 cup sliced green pepper
¼ small onion diced
1 dash Worcestershire sauce
1/3 cup shredded American cheese
½ tsp. freshly ground pepper
½ tsp. seasoned salt
1 drop garlic extract
1/3 cup sliced mushroom (optional)
1 drop Tabasco per egg

Beat eggs, milk, parsley, salt, pepper and garlic with a **FORK** until well mixed. Place mixture in medium-hot omelette pan. When half-done to taste, add worms, celery, green pepper, onion, cheese and mushrooms. Complete cooking and serve immediately.

Robert J. Smith, Half Moon Bay, Cal.

★ ★ ★

CURRIED VER DE TERRE AND PEA SOUFFLE

In a saucepan heat one cup of milk and stir in ¼ cup of grated fresh coconut and ½ teaspoon brown sugar. Let the mixture cool.

In ¼ cup butter sauce saute lightly one small onion, grated. Add one clove of chopped garlic and cook. Stir in 1½ teaspoons curry powder and gradually add the coconut mixture. Cook the entire mixture for about 10 minutes more, then add 3 tablespoons flour mix to a paste with milk. Cook five more minutes, and allow to cool slightly.

In another pan beat 4 egg yolks well and add curry sauce gradually while stirring. Mix in one cup of drained small cooked peas and one cup of prepared ver de terre cut in ½ inch lengths. Fold in 4 stiffly beaten egg whites, and season with salt and pepper.

Turn mixture into a buttered soufflé dish and bake soufflé in a 375 oven for 40 minutes.

George H. Lewis, Stockton, Cal.

★ ★ ★

VER DE TERRE IN DEVILED SHRIMP HORS D'OEUVRES

1.) Wash 1 cup earthworms, boil for 15 minutes over mod. heat. Rinse, repeat boiling, rinse and pat dry.

2.) Chop 3/4 cup blanched almonds, mix with earthworms, spread on a cookie sheet, and toast in hot oven.

3.) Boil 6 eggs and mash with 2 to 4 tab. of mayonnaise, 1 cup cocktail shrimp-drained and rinsed, ½ cup finely chopped celery, 3/4 cup grated cheddar cheese, 1/8 tsp. onion salt, 1/8 tsp. garlic salt, 1/8 tsp. salt. (option—sliced stuffed olives).

4.) Mix earthworm almond mixture with shrimp egg mix. Spread on sour dough bread squares, bake in moderate oven (350°) for 15 minutes, top with cheese and bake until cheese melts.

Serve for appetizers or as a main dish of open face sandwiches.

Vicky Ash, Arroyo Grand, Cal.

★ ★ ★

VER DE TERRE STUFFED PEPPERS

1.) Wash 2¼ cups earthworms and boil 15 minutes-mod. heat. Rinse, repeat boiling, rinse and pat dry.

2.) Fry together with ½ lb. of lean hamburger, 1 large onion, finely chopped, 2 cloves garlic, 1 tsp. parsley, 1/8 tsp. pepper, ¼ tsp. salt and 2 small cans tomato sauce, 6 large mushrooms thinly sliced.

3.) While you fry hamburger earthworm mixture, boil 4 to 6 bell peppers for 15 minutes or until tender.

4.) Mix into earthworm mixture 1 pkg. of long grain, wild rice of cracked wheat. Stuff peppers and mixture and bake at 350° for 25 minutes. Top with cheddar cheese and bake for 5 more minutes.

5.) Serve with chilled cream of avocado soup and carrot raisin salad.

Vicky Ash, Arroyo Grande, Cal.

★ ★ ★

The winner was Applesauce Surprise Cake by Patricia Howell of Minnesota. Ms. Howell received a check for $500 from North American Bait Farms, as well as a great deal of publicity. Bon appetit!

★ ★ ★

REFERENCES

1. Taylor, R. L. **BUTTERFLIES IN MY STOMACH**, Woodbridge Press, Santa Barbara, 1975.
2. Edwards & Lofty, **BIOLOGY OF EARTHWORMS**, op. cit. pg. 197.
3. Ibid. pg. 148.
4. Laverack, M. S. **THE PHYSIOLOGY OF EARTHWORMS**, MacMillan Co., New York, 1963, 206 pgs.

PHOTOGRAPH 10.1–Copulating Redworms
Research into how to stimulate earthworms to be more procreative is one type of research needed.

CHAPTER ELEVEN
THE LATEST ON RAISING EARTHWORMS TODAY

In the two years since VOL. I of this series was completed, we have seen tremendous growth in the vermiculture (worm-raising) industry in all aspects. New markets have opened up with surprising speed in the gardening industry with major nursery firms carrying earthworms in their catalogs, and potting soil manufacturers beginning to package earthworm castings for national distribution. At the same time, the fishing bait market for worms has continued to expand, and continued to experience shortage conditions.

As these markets have grown, so have the number of growers involved in raising worms to supply those markets. It has been estimated that as many as 90,000 people are now involved in some phase of the vermiculture industry on either a full-time or a part-time basis. With this surge in the number of growers has come a move towards greater industry cooperation and organization, through such organizations as the American Farm Bureau Federation and its state and local affiliates.

Many of these developments have been covered in detail in previous chapters. But there is yet another form of growth we must explore: the growth in technology. Continued research and development work by the authors and others have led to significant improvements in feeding, harvesting, and packaging techniques and equipment, as well as to improvements in pest management and other phases of good vermiculture practice.

A. VERMICULTURE

We should, perhaps, take just a few words, to discuss the term "vermiculture", which is itself less than two years old. The word is a combination of two Latin words "Vermes" meaning worm, and "cultura" meaning growth knowledge. Thus "vermiculture" refers to the knowledge or science of growing worms; primarily earthworms. The term, started by a NABF Spokesman, has since been used by all major news media, many scientists, government officials, and others to refer to the worm industry. "Vermiculture" is part of a group of enterprises such as "apiculture" (beekeeping), "aquaculture" (fish farming) and others which make up "agriculture" (growing crops in soil). The wide acceptance of this new designation for our industry reflects a growing recognition of the potential importance and contributions represented by our friend the earthworm.

Mini-Farms

This news clipping tells about mini-worm farms. Another new development occurs.

Weeder's Digest

THE NEVADAN—Sunday, September 7, 1975
Adelene Bartlett

Worm Farms

Ant farms have long been a source of fun and information, but now we have another similar type of farm, filled with critters that can't bite. This is a new type of mini-worm farm.

The mini-worm farm is 4-tiered, with each tier of a non-toxic type of plastic except for the bottom tray, which is solid white plastic. This bottom tray is the only one that isn't rotated and contains gravel in one layer in which the water for the farm is poured when the top layers seem to be getting a mite dry.

Each of the clear plastic tiers is about the size of a large shoe box and about as wide, although a little more shallow. The top lid is also of solid white and fits closely on the top tier to help keep the moisture in.

With proper care, which certainly isn't complicated, earthworms will live from 7 to 15 years and stay quite productive. They will consume their weight in food every 24 hours. Reproduction is best when the worms are active and searching for food.

The whole outfit is very neat and can be kept in any room of the house without any odor or mess while the earthworms go about their business in absolute quiet. Never have heard an earthworm utter even a squeak. They are different from pets that meow, bark, chatter, cheep or squawk, and they never ask to be let in or out or taken for a walk.

The farm cannot be kept in bright sunshine. It can be kept in absolute darkness, but will thrive in light conditions. The box can be artificially lighted but [they] will reproduce better if not in strong light.

Feed the earthworms about four times a month, allotting about 2 ounces of food each time. Thoroughly mix into the peatmoss in the top tray. Each tray has fine peatmoss bedding which can be replenished at local nurseries.

Recommended food is what we have in our kitchens except for the birdcage droppings which are mentioned in instructions on care. Corn meal, flour, coffee grounds, and bread crumbs are also mentioned.

It's possible that fruit and vegetable scraps can be used after pureeing them in a blender. All should be mixed thoroughly in the bedding in the top tray to allow the moisture and bacteria to work on it, making it even more palatable for the earthworm smorgasbord.

For children, these little farms can prove to be a profitable venture. Earthworm castings, which the earthworms produce faithfully in the trays, are figured at $1.75 per pound and tray No. 1 should produce $75.00 per year in castings.

If not used for sale the castings are valuable for dumping into planters, pots and flower beds as a plant food and also for the egg capsules which it will contain.

PHOTOGRAPH 11.1—**They grow healthy plants.**

In flowerbeds, earthworms will continue to eat and enrich the soil, creating little tunnels all through and around the roots, and bringing food, water and oxygen to the plants. If food there is provided on the surface, such as compost, manures and kitchen scraps, the beds will eventually be full of black loam even as these little mini farms are.

As an interesting pastime these farms can be used to provide hours of entertainment for shut-ins, for children, or for just anyone. They can be kept in apartments, mobile homes, and such and with no one the wiser about the livestock being cared for there.

As noted before, there is absolutely no odor of any kind with these farms. They are neat and clean as a pin. All the activity going on there is contained so no messy littering occurs. Each tray has holes in the bottom and the worms travel to and fro inside.

Children should be taught when young not to be squeamish about such wonders of nature as earthworms. They should be instructed not to come unglued over bugs and should learn about their place in the scheme of things.

As they watch the activities of these earthworms and learn all about them from the pamphlets which come with the farms, or from library material, it naturally leads to an appreciation of fellow creatures both large and small.

Who knows? Working with a worm farm like this might bring us a whole raft of budding earth scientists who might not otherwise have pursued such important work.

PHOTOGRAPH 11.2—Worm castings

A book on earthworms called "Earthworms for Ecology and Profit" published by North American Bait Farms, Inc. is available. This and the mini farms can be obtained by inquiring at Las Vegas Worm Farm, 7760 Haven St., Las Vegas, 39119 or phone 736-6681 or from NABF.

B. EARTHWORM FEEDING

The fact that earthworms will consume dairy manure as a primary food source has been one of the chief factors in encouraging the growth of the industry, because this manure is both cheap and readily available. However, as the cost of feed grains has gone up, dairymen have begun to reduce the protein value of the feed supplied to their herds, and this has reduced the protein value of the manure for worms.

In addition to the protein value, which should range between 12% and 16% (manure) for maximum worm production, there are other factors which have led wormgrowers to find supplements an increasing necessity for maintaining good production. The most common supplement used has traditionally been walnut meal. But walnut meal, largely as a result of its widespread use by worm farmers, has become increasingly scarce and unreliable, and increasingly expensive during the past two years. This has spurred the search for other materials.

Now, researchers at NABF and a major feed company working with computer analyses of scientifically determined nutritional needs of the earthworm, have developed a feed especially designed to provide all of the elements that earthworms require for maximum weight gain and breeding activity. This supplement is now on the market throughout the country, under the brand name WORM-GRO.®

WORM-GRO is a specially blended formula of organic waste products, vitamins, minerals, and hormones. It has a guaranteed analysis of not less than 19% protein, and not less than 5% fat; with less than 30% fiber and less than 10% ash. This is the highest guaranteed analysis protein of any material designed for earthworm feeding supplements now on the market.

Because of its high protein content and other factors, growers can achieve excellent results by feeding as little as 3 lbs. per week per 24 sq. ft. in addition to the regular feed used. It is recommended that this be increased to not more than 6 lbs. per week per bin for two weeks just prior to harvesting. These amounts should be added once a week. For regular feed, use 3 lbs. per 24 sq. ft. per week. Prior to harvest, feed 3 lbs. two times per week for each bin.

WORM-GRO may be obtainable through local feed stores in your area; particularly if there are a number of growers in your community. If you are a contract grower, your worm wholesaler should be able to supply you. Or you may write to one of the following firms for information on obtaining WORM-GRO for your operation:

Atlantic States	Brandywine Valley Ecosystems 1173 Telegraph road West Chester, Pennsylvania 19380
MidWest	Mid-America Worm & Ecology Farm P.O. Box 82 Garnett, Kansas 66032
Nevada	Las Vegas Worm Farm 7760 Haven Street Las Vegas, Nevada 89119
Northwest	Pacific Northwest Bait & Ecology 2504 N. Pearl Centralia, Washington 98531
Southern California	McCune's Bait Farm 11866 Lakeside Avenue Lakeside, California 92040
Southwest States	Cherokee Strip Worm Ranch Route 2 Deer Creek, Oklahoma 74636

And other dealers are added all the time.

C. HARVESTING EARTHWORMS

The job of harvesting earthworms from their beds has always been the most time-consuming and difficult part of this business, particularly when one is harvesting bait-size worms. The development of the table harvesting method, described in VOL. I (chapter 7), was a significant improvement over earlier methods of hand picking. However, even this method required hours of work by skilled and experienced harvesters to obtain the best possible yield from the bins.

Fortunately, American inventiveness is famous for solving just this kind of problem. With the growth of our industry, particularly in the number of larger operations (over 100 beds) which have been established in the past few years, the development of a mechanical device to perform the task of separating worms from their bedding was only a matter of time. But, as often happens in the development of new technology, many inventors begin working on the problem at the same time; producing a wide range of machines, some of which are better than others.

As the owner of the nation's largest earthworm marketing organization, I have been asked to examine most of the machines which have been developed to harvest worms. These requests have come from inventors seeking my endorsement for their harvesters. However, I have not been willing to make such an endorsement until I found a machine which I felt would truly serve the needs of the vermiculture industry. Even when I received letters and requests from growers wanting to know which machine to buy, I waited.

Finally, I found a machine I could recommend with real enthusiasm. It is called the "Down to Worms Harvester" and it is manufactured by the D&T Harvesting Machine Company, 44640 Oakglenn, Hemet, California. If you will write to this company, they will send you complete information, prices, and the name of their dealer nearest to your location.

The D&T harvester, shown in the pictures 11.3 and 11.4, has four adjustable legs, so that it can be set up easily on uneven ground. Most of the machines now on the market are too wide to roll between closely spaced beds; and a machine that can't get to the beds means much extra work for the grower. But the D&T Harvester, only 28 inches wide, rolls

PHOTOGRAPH 11.3—The D. & T. Harvester works quickly and easily wherever you want it.

easily between the bed rows. The D&T Harvester is easily adjusted to compensate for belt stretching and different bed conditions; and its rigid frame construction will handle any reasonable load.

With the D&T Harvester, weather conditions have no effect on harvesting operations, because, unlike most other harvesters, you can place the full depth of bedding onto the machine. The double screening system sorts worms from bedrun, returning the smaller worms to the beds. The machine can be easily operated by one man (other types sometimes require two people or more), and can be rolled quickly from one bed to another.

The D&T Harvester will, I believe, open a new era for the worm industry. Now growers can harvest larger quantities of worms, making their bed yields more efficient. Harvesting can be accomplished in much less time than by hand, and the worms are cleaner when harvested by this machine. The best feature about the machine is it does not matter the time of day *or* how wet your worm culture is: the machine will harvest them.

D. PACKAGING AND SHIPPING EARTHWORMS

With the growth of the vermiculture industry into a nationwide business, more and more orders involve shipment of fairly large quantities of earthworms long distances by air as well as by United Parcel Service. This created a need for specially designed packaging able to withstand the rough treatment often encountered during transit. Several container manufacturers have gotten into this business, but the best of them in my opinion is:

Peter Hewitt
Precision Paperbox
13563 E. Freeway Drive
Santa Fe Springs, CA 90670

This company makes shipping containers specifically for earthworms in several sizes as shown in the pictures above. You can write to them for prices and delivery times. These boxes are double walled, 200 lbs. test strength cardboard, wax coated inside, and vented. The boxes are preprinted, showing that their contents are live earthworms, and providing special handling instructions.

PHOTOGRAPH 11.4—These cartons can protect your earthworms.

Air Freight

If you are running a large operation and making numerous shipments by air, you should contact the air freight sales representatives from one or more airlines serving your area. Explain that you expect to ship substantial quantities of freight (worms), which will require special handling. Invite the sales representative to your farm to examine your product and packaging. Explain that your choice of carrier will depend on the carrier's ability to provide the special handling required.

Building a good relationship with a major airline will pay you great dividends in faster shipments and lower mortality rates. Many shipments of earthworms have been lost due to careless handling by freight personnel, particularly when boxes must be transferred from one plane to another before reaching their final destination.

E. PEST CONTROL

1. Help from Government Agencies

While the pests listed in VOL. I of this series are the most common faced by growers around the country, there are a number of other pests which may affect beds in certain areas, or at certain times of the year. One of the best sources of knowledge about how to handle these localized pests is your County Agent, or Cooperative Extension Service office.

Every county in the United States is served by at least one county agent. Usually his office address will be listed in your local telephone white pages in the directory serving your county seat. Most large farmers will also know how to reach the County Agent, as will the local office of your Farm Bureau, or your local Chamber of Commerce.

The County Agent can help you identify the particular insect pest troubling you, and can recommend control measures; both biological and chemical. Knowing your County Agent is a very good idea even if you do not have an immediate problem, as this individual is usually very influential with farmers and gardeners who may be customers for your worms.

PHOTOGRAPH 11.5—Japanese Beetle "grubs"—larvae

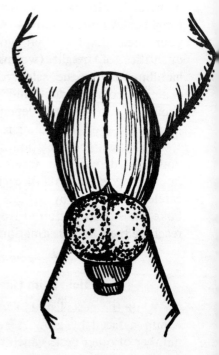

FIGURE 11.6—Japanese Beetle: color—body—metallic dark green; head and collar—shiny black; legs—black

2. Japanese Beetle

One pest that has recently become common in earthworm beds is the infamous Japanese Beetle. These beetles thrive in the rich soil of a properly maintained bed. While they are not predators of earthworms, they can do considerable damage to the structure of the bedding material, by mixing it with sand from the soil beneath the bed. The pictures shown are of a Japanese beetle grub.

The best control for Japanese beetles has been found to be a micro-organism known as milky spore disease. This organism, when added to the bedding, will infect the grubs, turning their blood a milky white and killing them before they become mature beetles. The spore powder will not harm earthworms in the beds, and only one application is necessary for permanent control, unless you remove all of the bedding and replace it entirely with new material. All that is necessary for control is to apply 2 teaspoonfuls of the powder to each bed. Apply the powder in two spots, in the center of the bed, about three feet from each end.

F. MACHINES HELP

Another new development comes from "down under." or the Southern Hemisphere.

MACHINE SPEEDS
THE JOB OF
EARTHWORM INTRODUCTION

by G. A. MARTIN Assistant Engineer, Ministry of Agriculture and Fisheries, N.Z.A.E.I., Lincoln College, and S. M. J. STOCKDILL, Farm Advisory Officer, Ministry of Agriculture and Fisheries, Palmerston.

For many years, the beneficial effects earthworms have on soil structure and subsequent pasture production have been recognized. Unfortunately, the beneficial species of earthworm *Allolobophora caliginosa,* and a few of lesser importance do not naturally occur in most undeveloped soils and are slow migrators. Consequently, a large proportion of land developed during the past 30 years, or longer in some areas, is sacrificing production due to their absence.

Although a satisfactory method of introduction has been devised, this has suffered from a high labor requirement, and many farmers are reluctant to adopt the practice. A machine, developed by the authors to cut blocks of turf, has substantially reduced the cost and labor requirement. This, we hope, will make the introduction of earthworms a much more attractive proposition to many farmers.

The presence of earthworms can lift pasture production by between 25 and 30 per cent, and can raise carrying capacity by 2.5 stock units per hectare. This is because the worms facilitate the deeper incorporation of plant nutrients; aid the breakdown and incorporation of organic material from the surface layer; transform compact, structureless soils to an open, friable condition; increase infiltration rate by as much as 100 per cent (important on country subject to soil erosion); increase the moisture-holding capacity of the soil; and promote greater proliferation and depth of plant roots.

On the debit side, earthworms have some less-desirable effects: Increased infiltration can increase soil leaching, so that slightly higher levels of maintenance liming may be required; soil cast on the surface may be eaten by stock with a consequent increase in teeth wear; with the removal of the organic surface mats and structural changes, soils with earthworms may be more susceptible to pugging.

These factors are generally insignificant when compared

PHOTOGRAPH 11.7—The turf-cutting machine attached to a tractor. The bladed wheel makes the transverse cuts, plough skeiths the longitudinal cuts, and a horizontal blade lifts the turfs.

with the advantages, and must be accepted as normal characteristics of high-producing soils.

Several methods of introducing earthworms to unpopulated areas have been tried in the past, but the one which has gained the greatest degree of acceptance involves cutting square spade turfs, 75 mm deep, from an earthworm-rich source area and laying these down on a ten-metre grid on the unpopulated area. This should be done when recent rain has drawn the worm population to the surface and when a dry or freezing-cold period after introduction is unlikely. This generally limits introduction to early spring or autumn.

Previously, turfs were hand-cut with a spade, but this job can now be done by the tractor-drawn machine illustrated here. It consists of a bladed wheel for transverse cuts, plough skeiths for longitudinal cuts and a horizontal blade for lifting out. The machine may be mounted on the three-point linkage of a medium-size tractor, and operating speed can be up to 8 km/hr. Source areas should be well grazed to allow constant machine working depth and although a limited number of stones in the soil can be tolerated, its use in excessively stony ground is not recommended.

Originally, it was intended that an elevator be fitted to the rear of the machine to facilitate the loading of a truck or trailer, but, because the expense of this would be considerable

PHOTOGRAPH 11.8—A machine-cut turf placed on a patch of limed pasture.

and the effort required for handloading is minimal, this has not been developed further.

Earthworms of the beneficial species require high calcium levels and moist soil conditions to stimulate their activity and migration.

Placing of turfs containing the earthworms involves applying 0.5 kg of lime to an area of one square metre at each point on a ten-metre grid and laying the turfs, grass side down, in the middle of the limed areas. The lime encourages initial development, and after a period of three to four years, when the worms have populated the area close to the blocks, a broadcast application of lime will encourage their further spread in soils where calcium levels would otherwise be too low.

Using the machine, three men can lift and lay turfs on 15 hectares in a day, without too much effort. One person spreads the lime, another places the turfs and the third drives the vehicle. Even with a 15 km transport distance for the turfs, the cost per hectare should be not more than $7, more than half of which would be for labor.

A machine for the automatic spreading of lime and placement of turfs is under development. This, it is hoped, will reduce the number of men required for distribution to two, and increase the speed at which it can be done.

PHOTOGRAPH 11.9—The turfs containing the worms have been placed at intervals of ten meters, in rows ten meters apart.

Although a single farmer may not be able to justify owning a turf-cutting machine for only a few days work a year, we feel that several farmers could combine to purchase or manufacture one. The mechanics of the machine are relatively simple, and any well-equipped farm workshop should have no difficulty in its manufacture.

Large numbers of these machines are not required and, as commercial production is not warranted, we have made available a standard construction plan. Copies of this plan may be obtained by writing to Mr. G. A. Martin, New Zealand Agricultural Engineering Institute, Lincoln College, Canterbury.

New Zealand Journal of Agriculture, Jan. 1976

G. MORE INFORMATION COMING

The author is now completing two books on additional aspects of worm-raising, which will both be published in 1977. The first of these books will be called: **The Care and Handling of Nightcrawlers.** This book will cover all aspects of handling nightcrawlers, particularly *Canadian* or *Native nightcrawlers,* the *African* and the *Gray*; from how to pick to how to sell retail, to how best to store and ship nightcrawlers, from picking ground to fisherman's hook.

The second volume will cover new developments in the techniques of raising earthworms under very cold or very hot climate conditions. This book will gather experiences from many worm farmers in the northern and desert parts of the country, with complete details on how to obtain maximum production year-round from beds in these areas. It will also cover different feeds in different areas.

Both of these future volumes will be available from the same source where you purchased this book.

Why not send in your questions, suggestions and problems to be covered in future books to Bookworm Publishing Co.?

★   ★   ★

CHAPTER TWELVE

SOIL TYPES AND STRUCTURES

Before we can put our full knowledge of earthworms to its best advantage in garden and agricultural soils, we must FIRST understand the soil itself. This is important because the earthworms affect and may be affected themselves by 1.) soil structure, 2.) soil chemistry and 3.) soil biology. The next few chapters are designed to give the reader who is unfamiliar with soil science a basic understanding of the environment in which earthworms make their home and to which they make such valuable contributions.

A. WHAT IS SOIL?

If you ask ten people, "What is soil?", you are likely to get ten different replies. Some will even say, "Ring around the collar". This type of reply indicates that the person has been watching too many TV commercials and soil to them is just plain dirt and grime. The preceding was, of course, intended jokingly, but many people do think of soil with a variety of definitions.

To those who have studied soil, it is a world in itself. A world populated with both micro- and macroscopic animal and plant life facing the same trials and tribulations of reproduction, life, birth, survival and death that face each of us in this world. (See Figure 12-1) This is why soil is often "dead" soil which does not benefit your potted plants. Plants need the interaction of a "living" soil.

A common definition of soil may be found in Webster's New Collegiate Dictionary. It says (1) "Firm land; earth. (2) The upper layer of the earth which may be dug, plowed, etc., in which something may take root and grow. (3) A country or land; a region. (4) Any substance, medium, etc., in which something may take root and grow; as social discontent is the *soil* in which anarchy thrives." Definitions (3) and (4) are not for our topic, but they do serve to illustrate the wide usage of the word, soil.

1. Soil

Soil, to the horticulturist, biologist or everyday gardener is basically *the soft layer that covers most of the land areas of the earth in which plants can grow.*

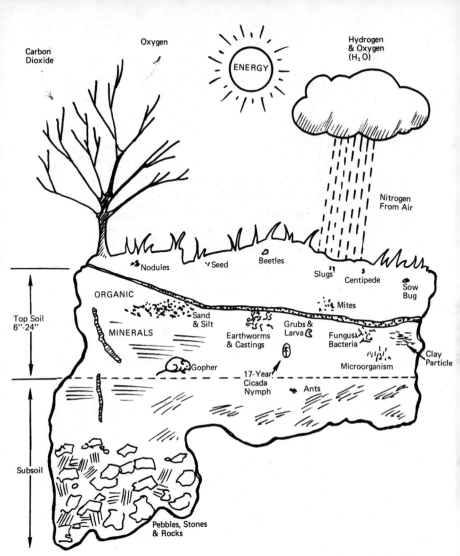

FIGURE 12.1—Soil: a phenomenon, a world within itself, a world populated with both micro and macroscopic animal and plant life facing the same trials and tribulations of reproduction, birth, life, survival, and death that man faces in this world and which we know so well.

Soil consists of 3 main ingredients or components as shown in Figure 12.2. These are:

a.) *Inorganic Material* which is basically rock granularized or reduced to grains and powder by the physical and chemical activities of the earth.

b.) *Organic Matter* i.) Dead or decaying organic material is a brown or black material formed by partial decomposition of vegetable, animal or other

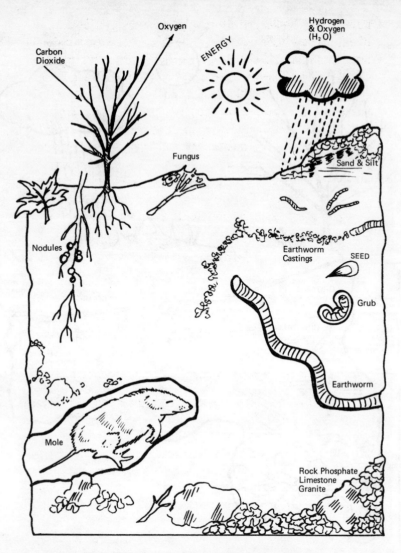

FIGURE 12.2–Soil biology and chemistry

organic residues in or on the soil. Various types of organic matter are listed in Chapter 15. The dead organic matter in the soil is known as *humus*.

ii.) Living organic matter in the form of micro- and macroscopic plants and animals. The microscopic plants are known as *microflora* and the microscopic animals as *microfauna* and their active life is classified as microbial activity. These microbes essentially consume decaying organic matter, convert it to humus and make nutrients and minerals available to plants. When microbes die, they break down to basic nutrients that plants can absorb or other microbes can consume. Many of the larger soil animals have the capacity for burrowing which provides the material necessary for microbial activity. Most of the larger soil animals, especially the earthworm, assist in the breakdown or fragmentation of the more decay resistant organic matter so that it is easier for the microbes to further decompose it.

and c.) *Soil Solution* which is a combination of soluble nutrients and minerals, water and dissolved gases.

2. Other Factors

A soil to be fertile also needs proper aeration or open spaces for air or atmosphere to penetrate to the plant roots. Two important factors that affect a soil's ability to support healthy plant life are its *temperature* and acidity or alkalinity known as the *pH* of the soil.

3. Inorganic Matter, Organic Matter and Humus

As with any complex subject, the preceding definitions and diagrams raise further questions such as "What is inorganic matter, organic matter and humus?" Again referring to Webster's New Collegiate Dictionary, two definitions are found for organic and inorganic, one for biology and for chemistry. Biologically, Webster defines *organic* as "Pertaining to, or derived from, living organisms.", *inorganic* as "Not organic; specifically, (a) Designating, or composed of, matter other than animal or vegetable; hence, inanimate. (b) Not forming, or not characteristic of, an organism." *Humus* is the decaying or dead organic matter in the soil.

B. SOIL FORMATION

In order that you may understand what these terms have to do with the soil, you must first learn how soil was originally formed. It is a very slow and still developing process all around the world.

Throughout this period spanning millions of years, many different types of rocks were formed: *igneous* rocks, such as granite, from the solidification of molten materials; *sediment-*

ary rocks, such as sandstone and limestone, from deposits of sediment (mud) which was later compressed or subjected to high pressure; and *metamorphic* rocks, such as marble and slate, from subjecting one of the other two basic types to great pressure and/or heat changing it to a more compact or crystalline form. These rocks are all classified as inorganic (non-living) and are composed of different minerals as a result of the varying chemical compounds available at the time they were formed.

Since its beginning, the earth has been in a constant state of upheaval with differing climatic and atmospheric conditions. Throughout the millions of years since their original formation, the various types of inorganic rocks have been broken up into boulders, smaller rocks and pebbles by the upheaval of the earth and sudden temperature changes, and then pulverized and granularized by the physical effects of wind, rain, water, ice and gravity. The rocks were also chemically eroded by the actions of acids formed by rain or other water mixing with the chemical compounds in the atmosphere or in the rocks.

Some soil forming processes other than by geological means are mentioned now. Some processes that produce soils of remarkedly varying types are:

1.) Calcification—the leaching and redistribution of calcium carbonate through the soil layers,

2.) Podzolization—the accumulation of organic material on the surface and the downward movement of colloidal solution clays to form a lower clay layer,

3.) Laterization—the decomposition of mineral materials to aluminum and iron oxides with the removal of silica,

4.) Gleization—the accumulation of wet or moist organic material,

5.) Salinization—the building up of salts,

6.) Alkalization—the building up of alkalies,

7.) Desalinization—the leaching or dispersal of such salts,

and 8.) Dealkalization—the leaching or dispersal of alkalies.

Gradually to our forming world of rock, air and water came organic compounds, microscopic life forms and then visual plants and animals. Organic life forms also aided the formation of soil. The end result of this physical and chemical reduction was a granularized or powdered material which we today know as soil. This is not a process which occurred once in the past and is now over; it is a continuing process.

C. TOPSOIL/SUBSOIL

The upper layer of the soil in which most plants grow and which contains the most humus or organic matter, microorganisms and small animals is called the *topsoil* as shown in Figure 12-1. Lower layers of soil which contain larger grains of rock and pebbles are called *subsoil*. Subsoil reaches down to bedrock, the solid rock of the earth's crust.

Our primary interest is with the topsoil where most plants grow. In some geographical areas, topsoil has been estimated to extend from 1 to 5 feet in depth. Chinese yellow or loess soil is a record of dozens of feet thick in which caves have been dug. However, thick topsoil cover generally only occurs in river valleys, on low-lying foothills, on the great plains and around lakes. In desert areas, there is practically no topsoil. In mountainous areas, there is generally only a thin layer of topsoil due to poor soil management and the eroding effects of wind, rain and gravity. The average depth of topsoil is normally 6 to 24 inches and it has been estimated that it takes nature from 500 to 1,000 years to produce 1 inch of topsoil.

Soil Color

A quick and simple (but not 100% reliable) soil test is looking at its color. Generally, but not always, the darker a soil is the more fertile it is. Humus content is what makes most soils *dark* and gives them their higher fertility than light coloured soils. However, some soils owe their dark brown or black coloring to their mineral components or to excessive dampness and these conditions are usually not of high fertility.

Soil that is *red* or red-brown in color can contain very large percentages of iron oxides if the local parent rocks were high in iron content and not subjected to excessive moisture. A red soil is generally a sign of a well-drained, not too humid and fertile soil. This is the case with the red soils of south-eastern United States. In other parts, the red color of the soil may mean that there are newly formed mineral materials that are not available for plant absorption.

Almost all *yellow* or yellowish soils are poor agricultural soils. Their yellow color is due to water reacting with iron oxides and thus is an indication that the land is now poorly drained or has been so in the past.

Grey or greyish soils can also be poor growing soils. The grey color can be a sign of a lack of iron or oxygen or that the soil has an excess of alkaline salts like calcium carbonate.

D. CAUSES OF A VARIANCE OF SOIL TYPES

The type of soil appearing in a given area will depend upon:

1.) The mineral content of the parent rock and the chemical compounds formed during the breakdown process. This determines the amount of such minerals as phosphorus, potassium, calcium, magnesium, sulfur, iron, manganese, boron, zinc, copper, molybdenum, and chlorine in the basic soil. This does not include minerals added to the soil by decaying organic matter—humus.

2.) The climatic conditions which existed during the time the rock was broken down. This determines the granularity or structure of the soil, the chemical compound formation and the depth of the topsoil layer.

3.) The type of plants and microscopic life (organic matter) which have been growing in the soil.

and 4.) The climatic conditions existing during the plant/microscopic growth and the length of time that growth has been going on. These factors—time and climate—then determine:

a.) the minerals in the soil,
b.) the depth of the topsoil,
c.) the granularity of the soil, which affects
its ability to hold moisture and create an atmosphere,
and d.) the number of microflora and microfauna in
the soil.

Thus it is an interdependent operation with one function depending on another. The climate determines which types of plants will grow in an area, and the type and abundance of microbial activity. Certain plants will only grow in specific climates. For instance, microbial activity increases greatly above 50 °F. Certain microbes thrive with little moisture while others require a great deal of moisture. Some plants extract far more of certain minerals from the soil than they return. The type of decaying plants and their mineral content help to determine the microbial activity and type because different microbes flourish where there is a concentration of particular minerals.

Microbial activity is necessary for plant decomposition and determines the rate of decomposition and the types of minerals which are returned to the soil. Some microbes flourish in an acid soil, extracting certain minerals while others flourish in alkaline soil and extract other minerals. The most productive microbes, *bacteria,* do best in a neutral soil.

In *summary* then, soil requires both organic and inorganic matter. Rock is broken down by nature to form the basic topsoil structure and then the basic topsoil is added to or 'built-up' by plants and micro-organisms as they die and decay. The plants, including trees, extract different minerals from various depths in the soil. When they die and decompose, these minerals are deposited on top of the soil. Various types of micro-organisms extract these minerals at a rate determined by the climate and soil acidity or alkalinity. The soil animals, such as earthworms, mix the decaying organic matter and minerals into the soil at various levels and help to stimulate microbial activity.

E. CLIMATE AND SOIL

The main soil areas of the world are closely related to the major climate zones. Six principal soils areas are:

1.) **Tundra**—has a very low average temperature, so very little microbial activity. Soils are shallow and underlaid with permafrost. Arctic soils support shallow-rooted mosses.

2.) **Podzols**—These soils are abundant in the cooler areas of the Temperate Zone. Formed in forest regions, podzols have been strongly weathered, have an acid or low pH and often are low in available plant nutrients. The top layer is often leached of valuable nutrients, but the lower layer can have more fertile mixtures of humus and clay.

3.) **Chernozemic**—These soils have developed under moderate rainfall under grass cover, such as the central plains. These chernozemic soils are often thick layers, rich in humus and plant nutrients as they have been subjected to little weathering. In temperate areas, they are among the world's most fertile soils, but in the tropics, the climate makes their management difficult.

4.) **Latosols**—These soils of the warm, humid lands are the most weathered soils of the world. Leaching has left them low in plant nutrients. Organic matter decomposes so fast that little organic content remains in the soil, but latosol soils are porous and allow water and plant roots to penetrate easily.

5.) **Desert Soils**—Hot, dry areas have desert soils which have had little leaching. Although they are low in humus, they are often high in inorganic nutrients.

and 6.) **Mountain Soils**—As temperature and rainfall varty at different altitudes on each mountainous area, so the vegetation and soils vary. A mountain could have latosol soil or desert soil at its base but probably tundra soil at its high peak. Many sub-types of soil are found in each main area.

F. SOIL TYPES

Scientists have identified thousands of different soil types, however, the three basic classifications are CLAYS, SANDS and LOAMS. Individual particles of a soil can range in size from a 2 inch stone to clay bits of 1/100,000 of an inch in length.

1. Clays

Clay soils consist of microscopic particles of alumin (aluminum oxide) and silica compressed into thin, flat layers. These layers, so small they are invisible to the naked eye, are relatively uniform and fit tightly together (less than .002 mm. between them) to form a compacted soil. A clay soil has usually undergone extensive chemical changes from the parent rock. The layers of alumina and silica have a large surface area, in comparison to their volume, which has a negative charge. Since opposite charges attract, this negatively charged clay soil can attract and hold positively charged minerals essential for plant growth. Thus fine clays provide a reservoir of nutrients for plants.

Since the particles fit so tightly together, clay soil is very slow to absorb water causing runoff and soil erosion. But the water that is absorbed is bound tightly in the clay particles, resulting in the soil becoming waterlogged after heavy rainfall or watering because it has poor drainage and aeration. Since clay soils retain moisture and have poor air circulation, they are slow to warm up in springtime.

Dry clay soil is almost impermeable to water or air and will harden to a thick crust. Clay soil is very hard to till but can be worked if not too wet. Most clay soils do contain the essential nutrients for plants, yet plants in clay soils can be stunted in growth because of lack of aeration to the roots and penetration by the roots.

2. Sands

Sandy soil is a loose, granular material which results from the disintegration of rocks and consists of particles of irregularly-shaped pieces 25 to 30 times larger (2 to .05 mm) than the clay particles. Usually, sand particles have not been chemically changed from the parent rock, so they can be rich in whatever minerals the parent rock had contained. These large, variably-shaped particles are loosely spaced which allows good air circulation, rapid or too rapid drainage and good tillage.

The loose spacing makes a too sandy soil a poor growing medium as the rapid drainage leaches out valuable minerals and nutrients as the water seeps to a level beyond the reach of plant roots. Sandy soil which has rapid drainage and free air circulation, generally warms up rapidly in the spring.

3. Silt

Silt, while not classified as a soil "type" proper, is a valuable element in a good soil. Silt is properly defined as a loose sedimentary material suspended in water, such as a deposit of sediment by a river. Silt is finer than sand; it consists of rock particles of .05 to .002 mm in diameter. Silt, like sand, usually retains the mineral characteristics of the parent rock.

These particles, smaller than ones of sand but larger than ones of clay, hold water better than sand and pack less tightly than clay. Dry silt is powdery like flour when sifted through the fingers, and it is greasy or sticky when wet. Silt can make soil slippery to the feel, but it will not harden like clay when dried.

4. Loams

Loam soils are ideal soils; they contain easily crumbled or pulverized mixtures of varying proportions of clay, sand, silt and a goodly percentage of humus or decayed organic matter. In nature, humus can be 2 to 5% of surface soil in humid regions, or less than .5% in desert areas, or up to 95% in peat bogs. If the loam is high in sand, it is known as sandy loam or if it is high in clay, then as a clay loam.

The clay, sand and silt particles in loamy soil are bound together into a granular or crumbly structural pattern over a period of years by the combined action of plant root growth, earthworms, microflora, microfauna and humus. Each granular crumb of a loamy soil holds a thin film of moisture which can be readily utilized by plant roots. This good granular structure allows excess water to drain off while retaining enough water for plant growth. The granular structure also allows free circulation of soil air and provides free space for good root growth. The higher the humus content the more nutrient availability, the better the water retention, air circulation and free space for root growth.

Loamy soil is usually dark and rich-looking. If a moist lump of loamy soil is picked up, it will retain its shape. If enough pressure is exerted, it can be crushed, but it will usually break up into smaller pieces which retain the granular structure of the parent clump. Loamy soil does not warm up as quickly as sandy soil in the spring, but it does warm up considerably faster than clay soils.

5. Other Soil Types

There are hundreds of different mixtures or sub-types of soils. Other soil types due to special conditions include *stoney* soils, *fill* soils and *wet* soils. These are NOT classified as soil "types" proper, they are kinds of soils which are likely to be encountered by anyone interested in soils, gardening or farming. One farm can have, at least, six different types of soil, so each area or field should be tested to see if it needs to be treated differently from the others.

a. *Stoney* soils are those soils which contain a good percentage of large rocks and stones in the topsoil layer. These rocks and stones present a problem when attempting to cultivate the soil; therefore, they should be removed before cultivation begins.

b. *Fill* soils may be one of two types, both of which are generally found in areas of new construction. One type, normally brought in to level an area, may or may not be topsoil. In many cases, it is an infertile subsoil. The other type of fill soil contains a large percentage of debris such as lumber, large rocks, nails, glass, and wallboard. Either type of fill soil requires a great amount of work before it will be productive. It is generally faster and less expensive to remove the top 8 to 12 inches and replace it with a good topsoil.

c. *Wet* soils or swampy areas are generally clay soils or those soils in areas with a high underground water table. This type of soil will remain puddled for a long period of time after a hard rain, or will remain wet in the spring when other soils are only moist. If the soil is clay, better drainage can be provided by adding sand or organic matter to a depth of 18 to 24 inches. If the wetness is caused by a high water table, it may be necessary to pump the water out or, if possible, install under ground drain pipes to a run-off area. Most plants cannot stand a wet soil for long periods of time; therefore, the condition must be corrected before attempting to plant the area.

But remember, no one soil type is "ideal" for all plants. Rice and sugar cane want far more moisture in the soil than do wheat and corn crops.

G. SOIL STRUCTURE

The various types of soil have been discussed without regard to the overall structure of the soil necessary for ideal plant growth. Various factors affect soil structure which in turn affects plant growth rates.

Soil structure refers to the overall proportion of components in the soil. An ideal ratio of components in the soil with a good composition is shown in Figure 12.3. A soil which contains 50% solid matter (mineral and organic) and 50% pore space is considered as the optimum for plant growth. Approximately ½ of the pore space should be filled with moisture and the other ½ soil air. Mineral matter should be nearly 45% and organic matter the remaining 5% of the soil. This is a good soil mixture as it allows water movement, gas exchange, microbial activity and provides loosely spaced particles for plant root growth.

Very few soils are so balanced; most are either too much

clay or too much sand, providing a less than optimum balance between air and water in the pore spaces and the solid matter. Clay soils retain water but not enough air, and sandy soils are airy but water drains away too rapidly.

Decomposing organic matter, or humus, in the soil will improve the overall soil structure because the humus combines with the various size soil particles to form a more granular structure. However, most soils do not contain enough organic matter and the amount of organic matter in the soil is constantly being lessened by decomposition. A

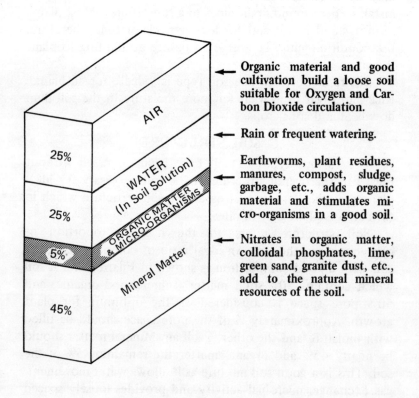

FIGURE 12.3—Soil structure

continual addition of organic matter and earthworms to the soil will help to maintain the soil closer to the optimum condition.

Soil structure and plant growth are also affected by environmental factors such as moisture, temperature, aeration, fertility, and soil acidity or alkalinity. Most of these factors can be partially controlled by addition of organic matter, or humus, to the soil.

1. Moisture

Moisture is one of the more important components in soil. It performs the following function with relation to building soil structure and assisting plant growth:

a. Moisture provides some oxygen which is taken in by plant roots. In a soil which is too wet, however, there is little aeration and plants may be affected by a lack of oxygen and a build up of toxic carbon dioxide.

b. Moisture dissolves the nutrients and minerals in the soil and makes them available for absorption by plant roots.

c. Moisture affects and influences the decomposition of dead organic matter. The most beneficial groups of microbes are highly active in a soil with an optimum moisture condition. In a soil which is too dry, there is little, if any, microbial activity. In a soil which is too wet, undesirable microbial activity may occur. In the latter case, anaerobic microbes, which convert nitrates into gaseous nitrogen, sulfates into sulfides, and deplete the soil of oxygen, may become predominant. (Anaerobic microbes are those which live and are active in the absence of oxygen.) Additionally, many plants do not grow well in an overly wet soil because of this undesirable microbial activity.

Soil moisture can be controlled somewhat by the addition of an organic mulch cover to the soil. Such a cover usually causes the soil to maintain a higher and more constant moisture level. Even with good cultivation, most soils will tend to pack and crust after a heavy rain or watering. The addition of a layer of loose porous material such as grass clippings, hay, sawdust, straw, or wood chips on top of the soil acts as a cushion for the water. This allows water to seep slowly into the soil without beating directly on the soil and causing it to become compacted. Additionally, a mulch prevents the loss of moisture by reducing the rate of evaporation from the soil.

As you may recall from reading Chapter 2, earthworms have been proven to be an aid in the soil's ability to absorb and retain moisture.

2. Temperature

Temperature plays an important part in the soil environment. It affects the timing of plant growth, partially determines the degree of microbial activity, and in many cases, determines plant survival. As noted in the discussion on soil types, sand soils tend to warm up fast in the spring, while the heavy clay soils, which retain moisture and admit very little air, tend to warm up slowly. The earlier warmth of a sandy soil means earlier plant growth. The large pore spaces and good aeration of the light sand soils favor a balance between soil temperature and the atmosphere. Organic matter added to either sandy or clayey soil will improve aggregation or granulation, especially in the clay soils which will then respond more rapidly to temperature changes.

In temperature below 50 °F, microbial activity begins to slow down and in temperatures below freezing, practically stops. In the springtime, after soil temperatures reach 50 to 60 °F, microbial activity starts to increase. A temperature of 85 to 90 °F is the optimum temperature for the high state of microbial activity necessary for plant decomposition. Microbial activity will decrease with higher temperatures and will practically stop at temperatures above 100 °F.

An organic mulch may be used to increase the soil temperature in winter or decrease the soil temperature in summer. This occurs as a mulch is slower to change temperature than a mineral soil, so it is less affected by day and night temperature changes.

3. Aeration

Aeration is another important factor which affects the soil structure and the soil environment necessary for microbial activity and plant growth. Aeration refers to the looseness of the soil and the space between the solid matter particles. Soil aeration can be improved by cultivation and by the addition of organic matter or humus, to the soil.

Cultivation will loosen the soil, providing better circulation. The granular structural soil improvement brought about by the addition of organic matter will help to maintain the soil in a loose and porous condition. Heavy farm machinery can pack surface soil tightly.

In a soil which is properly aerated, air (oxygen, carbon dioxide, and other gaseous compounds) can freely circulate throughout the soil. This "soil air" provides a constant supply of oxygen and carbon dioxide to the roots of plants, and to the micro-organisms in the soil, and allows dispersion of the carbon dioxide given off by these roots and micro-organisms into the atmosphere. Additionally, a well-aerated soil will generally support the microbial activity necessary to convert available nutrients into readily absorbed plant food. Conversely, in a soil which is not properly aerated, the microbial activity is generally reduced and a non-beneficial group of microbes active. These microbes not only compete for the available oxygen supply and take away the oxygen necessary for plant growth, but also can convert oxidized compounds, such as nitrates and sulfur, into a form which is not usable by plants.

A soil which is not properly aerated can also become compacted and hold too much moisture. In addition to the previously listed adverse effects of too much moisture such a condition can prevent air from moving through the soil, causing a lack of oxygen necessary for plant growth and microbial activity, and allowing a severe build up of carbon dioxide which is dangerous to the micro-organisms in the soil.

4. Acidity and Alkalinity (soil pH)

Acidity or alkalinity (soil pH) is another important factor affecting microbial activity and plant growth in the soil. If the soil is too high in acid content, one group of microbes becomes active. If it is too high in alkaline content, another group of microbes becomes predominant. Generally, fungi are more active in an acid soil and bacteria (including aztobacter which fixes nitrogen) become inactive. In an overly alkaline soil, nutrients and minerals necessary for plant growth are not produced by microbial activity. The soil becomes what is commonly known as infertile. This is true of many barren desert areas which have a high alkaline content.

A nearly neutral soil is best for microbial activity and plant growth. Soil pH and methods of controlling it are discussed in Chapter 13.

5. Soil Fertility

Soil fertility is not only affected by the mineral content of the soil, but also by the soil type and structure. Many of the valuable nutrients in the soil which are necessary for plant growth are held in a water film which surrounds each soil particle. These nutrients are held in a "soil solution" from which they are assimilated by the plant roots. Generally, the more granular a soil, the more nutrients which are available.

As explained in the discussion on soil types, the fine, tightly-packed particles in clay soils provide more surface area to which nutrients may cling, but they lack the granular structure needed for an atmosphere. Sandy soils are so loose that the nutrients are leached below the level reached by plant roots. Organic matter added to either type of soil can not only improve the soil as previously noted, it can help to increase the fertility of the soil. The addition of organic matter to clay soils causes the clay particles to become aggregated so that plant roots can more easily penetrate the soil and come in contact with the soil solution which contains plant-growth nutrients. Organic matter added to sandy soil holds the nutrient-containing soil solution in the upper soil level where it is available to the plant roots. Earthworms always increase the benefits of added organic matter to your soils.

H. IMPROVING SOIL STRUCTURE

Except for the smallest planters and areas, it is impossible to build a good, rich, loamy soil by simply mixing the component parts. However, as previously noted, almost any soil can have its air circulation, moisture holding ability or drainage, and mineral content improved by adding organic matter such as compost, leaf mold, sawdust, ground bark, peat moss, or manure. The end result, with proper addition of these components, will be a rich, crumbly, granular soil. Organic matter for soil enrichment is discussed in detail in Chapter 15.

Drainage and aeration of clay soils can be improved by the addition of either gypsum or lime, each of which contains calcium. The calcium will cause the clay particles to cling together, forming soil crumbs which are larger than the individual clay particles. This, in turn, provides larger air spaces between the crumbs. Gypsum and lime do not contain nutrients and do not directly aid the soil micro-organisms; therefore, they should not be used as a substitute, but only in addition to, organic matter. Gypsum should be used if the clay soil is highly alkaline, and lime if the soil has a high acid content.

We know the background information of how organic matter improves both the too fast drainage of too sandy soils and the too slow drainage of too clayey soils.

The modern, successful gardener and farmer base their soil management on solving five basic problems:

    1.) proper tillage,
    2.) maintenance of a proper nutrient supply,
    3.) keeping enough organic matter in the soil,
    4.) maintenance of a correct soil pH,
and 5.) control of erosion.

Remember you can consult government agricultural departments and/or your local agricultural colleges about specific soil problems.

★ ★ ★

CHAPTER THIRTEEN

PLANT GROWTH AND SOIL pH

Before discussing the nutrient content of soil, or soil chemistry, it is probably a good idea to clarify a popular misconception; that is, you must "feed" your plants "XKY" fertilizer or "I Make Um Grow" plant food. Your plants cannot eat; therefore, you cannot feed them food. This misconception is caused by the high-powered advertising put out by fertilizer manufacturers to sell their product.

All plants with roots, from a blade of grass to the tallest tree, require the same basic elements for growth (Figure 13-1). They need: light, water, warmth, and air above ground, and air, water, warmth, and nutrients below ground. Nutrients in the soil are not plant "foods" regardless of the advertising misconception. Plants, unlike animals, manufacture their own foods through processes called photosynthesis and respiration.

Photosynthesis is a process which synthesizes (forms) chemical compounds from basic chemical elements or other compounds when exposed to light.

Respiration is a process whereby a plant absorbs oxygen from the air and gives off the products (especially carbon dioxide) formed by the oxidation of chemicals in the tissues. This *oxidation* process can be simply explained as the "burning" of sugars or carbohydrates which had been manufactured to obtain energy for growth.

Green plants do use some oxygen and do emit some carbon dioxide as do animals (including man) in respiration; *but* in their photosynthesis process plants use 5 times the amount of CO_2 they produce and they produce 5 times the amount of oxygen they use. Thus plants are necessary to produce our needed oxygen while removing much of the carbon dioxide from our atmosphere.

Plants take in energy from the sun, carbon dioxide from the air, and water from the soil to manufacture their own food. Sunlight on the foliage reacts, through photosynthesis, with the green chlorophyll, carbon (from carbon dioxide), and other elements in the leaves to make food for the plant

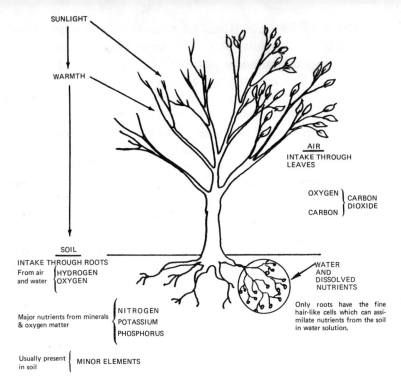

FIGURE 13.1—Plant nutrition

in the form of carbohydrates or sugars. The manufactured food moves down the stems to the root area where it helps to promote growth. The roots take in water and nutrients from the soil which are transported up the stem or trunk to the leaves as raw material for photosynthesis. This liquid material is what gives the leaves their substance; if it is lacking, the leaves wilt and die, and the whole process slows down. The manufactured food can supply immediate energy, be stored for future use, or be used as building material within the plant. Every part of the plant, leaves, roots, stems, etc., must have a constant supply of this energy ("food") to continue living.

While the food is being manufactured by the plant, an opposite process—respiration—is simultaneously taking place. As with the human body, the carbohydrates or sugars are being burned to supply energy to the plant. In this burning process, oxygen is taken into the plant, and carbon dioxide is released into the atmosphere and the soil along with the excess oxygen from the photosynthesis process.

The only way plants can take nutrients and water from the soil is through tiny, hair-like cells at the end of the root. These cells can assimilate nutrients only if they are in soil solution. This is why the space for air and moisture between the soil particles is of great importance.

There are approximately 16 known elements required for plant growth. Some of these elements are readily soluble while others must be released by the action of the billions of micro-organisms in the soil before they are available for assimilation by plants. Most soils have enough of the minor or "trace" elements; however, the three major elements: nitrogen (N), phosphorus (P) and potassium (K), are usually required in amounts larger than that available in most soils. Organic matter added to the soil not only improves the soil structure, but also helps to add the three required major elements.

Carbon dioxide, which is released during the energy "burning" process by the plant, is also taken in to synthesize the plant food. The carbon is combined with hydrogen to form carbohydrates or sugars, and the oxygen is used in the burning process. The soil micro-organisms, which break down organic materials, obtain energy in a similar manner. They use organic matter for food and, as they break it down, take in oxygen and release carbon dioxide. In this way, the soil, the plants, the micro-organisms and the atmosphere are all very closely bound one to the other in a living cycle.

B. SOIL pH

References have been made in the previous paragraphs to soil acidity and alkalinity. The degree of acidity or alkalinity is referred to as the soil pH, and is measured on a logarithmic scale. The soil pH may affect plant growth in various ways at both extremes of the scale. Some soils may be too acid (low pH) for optimum growth while others may be too alkaline (high pH).

Soil pH affects the microbial activity necessary to release or convert nutrients into usable forms for plants, and a plant's ability to assimilate or use those nutrients which are available. Experienced amateur gardeners and professional horticulturists both agree that one of the most important factors in growing plants is a properly regulated pH condition within the soil. Earthworms thrive in a neutral soil.

1. pH Defined

pH refers to the potential hydrogen and is a relative measure of the numbers of hydrogen ions (positively charged atoms) and oxygen-hydrogen (hydroxide) ions (negatively charged molecular units, each consisting of one hydrogen atom and one oxygen atom) in soil-water solution. If these two ions are present in equal numbers, the soil is neutral. If more hydrogen ions are present, the soil is *acid*. If more hydroxide ions are present, the soil is *alkaline*. Since water is a very stable compound, only a little of this ionization actually takes place. Nevertheless, soil acidity or alkalinity is determined by this ionization. It is caused by the reaction of various minerals and organic compounds with moisture in the soil.

The acid-alkalinity scale ranges from 0 to 14 as shown in Figure 13-2. The low end of the scale indicates acid, and the high end indicates alkaline. Seven is exactly neutral. The importance of a properly regulated pH can best be understood when it is realized that pH values are in multiples of 10. That is, a pH of 6 is 10 times more acid than a neutral pH of 7; while a pH of 5 is 100 times more acid, and a pH of 4 is 1000 times more acid than a neutral pH of 7. Furthermore, the addition of fertilizers and the continual leaching action of rain and watering cause a soil to become more and more acid in high rainfall areas. The soil in low rainfall areas is generally alkaline, and is more acid.

Soil pH, from a chemical standpoint, is quite complex. Soil particles contain negative electrostatic charges on their surfaces due to their chemical composition. This electrostatic attraction will attract and hold positively charged particles such as hydrogen ions. These positively charged hydrogen ions are exchangeable. This means that they may be replaced by other positively charged ions. When calcium carbonate or limestone is added to acid soil, the positively charged calcium ions replace the hydrogen ions. The hydrogen ions are then neutralized by the chemical reaction with the limestone (pH 7).

If the positively charged hydrogen ions were the only factor causing soil acidity, a small amount of limestone would be all that would be required to raise soil pH. However, due to the positively charged ion exchange capacity

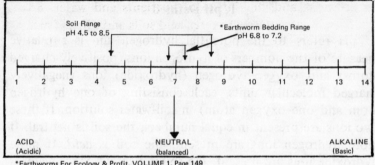

*Earthworms For Ecology & Profit, VOLUME 1. Page 149

See Chart 13.3. pH range of common fruits & vegetables and herbs.

FIGURE 13.2–The pH scale

of soils, they have a large resistance to change, and a large quantity of limestone must be added to effect any change in the soil pH. There are two groups of ions whose concentrations are constantly fluctuating with changes in soil pH; aluminum and hydrogen versus calcium, magnesium, and potassium. As the soil pH becomes more acid, there is a large increase in hydrogen ion concentration which reduces the availability of calcium, magnesium and potassium to plants.

2. Effects of pH on Plant Growth

a. The solubility and subsequently the availability to plants of nutrient elements is related to soil pH. Some nutrients are highly soluble at high (alkaline) pH values, while others are a more available or soluble at low (acid) pH values. The maximum availability of most plant nutrients occurs between pH 6.5 and 7.0.

b. Soil pH may directly affect root cells, causing a reduced permeability for intake of water and nutrients, resulting in poor growth. The protoplasm of the plant cells (exclusive of the nucleus) consists mostly of protein molecules which can be extensively altered by excessive hydrogen or hydrogen ions.

Most plants will grow fairly well over a range of values from about pH 5.5 to pH 7.5. However, each plant species has a specific pH value at which it generates optimum growth. This optimum growth is, of course, related to root development and the intake of nutrients and water. Some plants (such as potatoes) prefer acid soils and others (such as artichokes) prefer alkaline soils. Most vegetables and flowers, however, prefer soils that are neutral or slightly acid $pH\ 6.5$). The pH range of common fruits, vegetables, and herbs is listed in Table 13.3.

TABLE 13.3–Reference list of plants with range of pH preferences.

Plant	pH	Plant	pH	Plant	pH
African Daisy	6.5-7.5	Catalpa	6.0-8.0	Flytrap, Venus	4.0-5.0
African Violet	6.5-7.5	Cauliflower	6.0-7.0	Forgetmenot	6.0-8.0
Ageratum	6.0-7.0	Celery	6.0-6.5	Forsythia	6.0-8.0
Alder	5.5-6.5	Centaurea	6.0-8.0	Four-O-Clock	6.0-8.0
Alfalfa	6.0-7.0	Cerastium	6.0-7.0	Foxglove	6.0-8.0
Almond	6.0-8.0	Cherry	6.0-8.0	Franklinia	5.0-6.0
Alsike Clover	6.0-7.0	Chestnut, Amer.	5.0-6.0	Fuschia	6.5-7.5
Alyssum	6.0-8.0	Chicory	5.5-6.5	Gaillardia	6.0-8.0
Amaryllis	5.0-6.0	China-aster	6.0-8.0	Galax	4.0-5.0
Ampelopsis	6.0-8.0	Chive	6.0-7.0	Garbera	6.0-7.0
Anemone	6.0-8.0	Chrysanthemum	6.0-8.0	Garden Cress	6.0-8.0
Andromeda	4.5-5.5	Cineraria	7.0-8.0	Gardenia	5.5-6.5
Apple	5.5-6.5	Clarkia	6.0-6.5	Garlic	6.5-7.5
Arborvitae	6.0-8.0	Clematis	6.0-8.0	Genista	7.0-8.0
Arbutus, Trail.	4.0-5.0	Clethra	5.0-6.0	Geranium	7.0-8.0
Artichoke	7.0-8.0	Clover	6.0-7.0	Gerbera	7.0-8.0
Ash	6.0-8.0	Coffee	5.0-6.0	Gilia	6.0-8.0
Asparagus	6.0-7.0	Coleus	6.0-8.0	Ginkgo	6.0-8.0
Aster	6.0-8.0	Collards	6.0-8.0	Ginseng	6.0-8.0
Astilbe	6.0-8.0	Colorado Spruce	6.0-7.0	Gladiolus	6.0-8.0
Avocado	6.0-8.0	Columbine, Hyb.	6.0-7.0	Globeflower	6.0-8.0
Azalea	5.0-6.0	Coneflower	6.0-8.0	Godetia	6.5-7.5
Bachelor B'ton	6.5-7.5	Convolvulus	6.0-8.0	Gooseberry	5.5-6.5
Banana	7.0	Coral Bells	6.5-7.5	Grape	6.0-8.0
Barberry	6.0-8.0	Coreopsis	5.5-6.5	Grapefruit	5.0-7.0
Barley	6.0-7.0	Corn	6.0-7.0	Grass, Orchard	6.0-8.0
Bayberry	5.0-6.0	Cosmos	6.0-8.0	Grass, Bermuda	6.0-7.0
Beach Plum	6.0-8.0	Cotoneaster	6.0-8.0	Grass, Co. Bent	5.5-6.5
Bean	6.0-7.5	Cotton	5.5-6.5	Grass, Creep. Be.	6.5-7.5
Bean, Lima	5.5-6.5	Coxcomb	6.5-7.5	Grass, Ita. Rye	6.5-7.5
Beech	6.0-7.0	Crabapple	6.5-7.5	Grass, Peren. Rye	6.5-7.5
Beebalm	6.0-7.0	Cranberry	5.0-6.0	Grass, Rough Bl.	6.5-7.5
Beet	5.8-7.0	Cranesbill	6.0-8.0	Gypsophila	6.0-8.0
Beet, Sugar	6.0-8.0	Crocus	6.0-8.0	Hawthorne	6.0-8.0
Begonia	6.0-8.0	Cucumber	6.0-8.0	Hazelnut	6.0-7.0
Bellflower	6.0-8.0	Currant	6.0-8.0	Heath	5.0-6.0
Bentgrass	5.5-6.5	Cyclamen	6.0-8.0	Heather	5.0-6.0
Birch	5.0-6.0	Cypress	5.0-6.0	Heliotrope	6.0-8.0
Bittersweet	5.5-6.5	Daffodil	6.0-6.5	Hemlock	5.0-6.0
Blackberry	5.5-7.0	Dahlia	6.0-8.0	Hepatica	6.0-8.0
Blackcap	6.0-7.0	Dandelion	6.0-8.0	Hibiscus	6.0-8.0
Bleedingheart	5.0-6.0	Daphne, Winter	5.0-6.0	Hickory	6.5-7.5
Bloodroot	5.5-6.5	Daphne, Feb.	7.0-8.0	Holly, Inkberry	4.0-5.0
Bluebell, Va.	6.0-8.0	Daylily	6.0-8.0	Holly, Amer.	5.0-6.0
Blueberry	5.0-6.0	Delphinium	5.5-6.5	Hollyhock	6.0-8.0
Bluegrass	6.0-8.0	Deutzia	6.0-8.0	Honeylocust	6.0-8.0
Bouvardia	5.5-6.5	Didiscus	6.5-7.5	Honeysuckle	6.0-8.0
Boxwood	6.5-7.5	Dogwood, Flwg.	6.0-7.0	Hornbeam	6.0-8.0
Broccoli	6.0-7.0	Douglas-fir	6.0-7.0	Horseradish	6.0-8.0
Broom, Scotch	5.0-6.0	Dutchmans pipe	6.0-8.0	Huckleberry	5.0-6.0
Brussel Sprouts	6.5-7.5	Egg Plant	6.0-7.0	Hyacinth	6.0-8.0
Buckwheat	5.5-6.5	Elaeagnus	6.0-8.0	Hydrangea, Blue	4.5-5.5
Bugbane	5.0-6.0	Elder	6.0-8.0	Hydrangea, Pink	6.0-8.0
Burning Bush	5.5-6.5	Elm	6.0-8.0	Inkberry	5.0-6.0
Buttercup	6.0-8.0	Endive	6.5-7.5	Iris	6.0-8.0
Butterflybush	6.0-8.0	English Ivy	6.0-8.0	Iris, Japanese	5.0-6.0
Butterflyflower	6.0-8.0	Eucalyptus	6.0-8.0	Ivy, Boston	7.0-8.0
Cabbage	6.0-7.0	Euonymus	6.0-8.0	Ivy, English	7.0-8.0
Calendula	6.0-8.0	Euphorbia	5.5-6.5	Jack-in-pulpit	5.0-6.0
Calla	4.0-5.0	Even. Primrose	6.0-8.0	Jacob's Ladder	5.0-5.5
Camas	6.0-8.0	Fern, Asparagus	6.0-8.0	Juniper	5.5-7.0
Camelia	4.0-5.5	Fescue, Chewing	5.5-7.0	Kale	6.0-8.0
Candytuft	6.0-7.0	Feverfew	6.5-7.5	Kalmia	4.0-5.0
Canna	6.0-8.0	Fir	5.0-6.0	Ky. Coffee	6.0-8.0
Cantaloupe	6.0-8.0	Firethorn	6.0-8.0	Kerria	6.0-8.0
Carnation	6.0-8.0	Flax	6.0-7.0	Kohl-Rabi	6.5-7.5
Carrot	5.5-6.5	Flwg. Quince	6.0-8.0	Laburnum	6.0-8.0

(Continued on following page.)

(TABLE 13.3 continued:)

Ladyslip. pnk	4.0-5.0	Passionflower	6.0-8.0	Shallot	5.5-6.5
Larch	5.5-6.5	Pea	6.0-8.0	Shootingstar	6.0-8.0
Larkspur	6.0-8.0	Peach	6.0-8.0	Snapdragon	6.0-7.0
Leek	6.0-8.0	Peanut	5.0-6.0	Snowdrop	6.0-8.0
Lemon	5.5-7.0	Pear	6.0-8.0	Solomonseal	6.0-8.0
Lentil	5.5-6.5	Pecan	6.0-7.0	Sorghum	5.5-6.5
Lespedeza	5.5-6.5	Pentstemon	6.0-8.0	Sorrel	6.0-7.0
Lettuce	6.0-7.0	Peony	6.0-8.0	Soybean	6.0-7.0
Leucothoe	5.0-6.0	Pepper	6.0-6.5	Spicebush	5.5-6.5
Lilac	6.0-8.0	Petunia	6.0-8.0	Spiderwort	6.0-8.0
Lily	5.0-6.0	Phlox, Annual	6.0-8.0	Spinach	6.5-7.0
Lima Bean	6.5-7.5	Phlox, Creep.	5.0-6.0	Spinach, N. Zea.	5.5-7.0
Linden	6.0-8.0	Phlox, Garden	6.0-8.0	Spirea	6.0-8.0
Lobelia	6.0-8.0	Pine	5.0-6.0	Spruce	5.0-6.0
Locust	6.0-8.0	Pineapple	5.0-6.0	Squash	6.0-8.0
Loosestrife	6.0-8.0	Pink	6.0-8.0	Star-of-Beth.	6.0-8.0
Lupine	5.0-6.0	Planetree	6.0-8.0	Stock	6.0-7.0
LoosestrifePrp.	6.0-8.0	Plum	6.0-8.0	Strawberries	5.0-6.0
Magnolia	5.0-6.0	Poinsettia	6.0-8.0	Sunflower	6.0-8.0
Maple	6.0-8.0	Poplar	6.0-8.0	Sweet Peas	7.0-8.0
Marigold	6.0-8.0	Poppy	6.0-8.0	Sweet Wm.	7.0-8.0
Marjoram	6.0-8.0	Potato	4.8-6.5	Swiss Chard	6.5-7.5
Marshmarigold	6.0-8.0	Potato, Sweet	5.5-6.5	Tamarix	6.0-8.0
Mignonette	6.0-8.0	Pricklypear	5.0-6.0	Timothy	6.0-7.0
Millet	5.5-6.5	Primrose	6.0-8.0	Tobacco	5.5-7.5
Milo	6.5-7.5	Privet	6.0-8.0	Tomato	6.0-7.0
Mint	6.0-8.0	Pumpkin	5.0-5.5	Trillium	6.0-7.0
Mockorange	6.0-8.0	Pyrethrum	6.5-7.5	Trmpetcreeper	6.0-8.0
Monkshood	6.0-8.0	Quince	6.5-7.5	Tuberose	6.5-7.5
Morning-glory	6.0-8.0	Radish	6.0-8.0	Tulip	6.0-7.0
Mountain-ash	4.0-5.0	Raspberry	5.0-6.0	Tuliptree	6.0-7.0
Mt. Laurel	5.0-6.0	Redbud	6.0-8.0	Tupelo	6.0-7.0
Mulberry	6.0-8.0	Redcedar	6.0-7.0	Turnip	6.0-8.0
Mushroom	6.5-7.5	Red Clover	6.5-7.5	Verbena	6.0-8.0
Muskmelon	6.0-7.0	Red Hot Poker	6.5-7.5	Vetch	6.0-7.0
Myrtle	6.5-7.5	Redtop	6.0-7.0	Viburnum	6.0-8.0
Narcissus	6.0-8.0	Retinospora	6.0-8.0	Violet	6.0-8.0
Nasturtium	6.0-8.0	Rhodedendron	5.0-6.0	Walnut	6.0-8.0
Oak	6.0-7.0	Rhubarb	6.5-7.5	Watercress	6.0-8.0
Oats	6.0-7.0	Rice	6.0-7.0	Water Lily	5.5-6.5
Oconne-bells	5.0-6.0	Rose	6.0-8.0	Watermelon	6.0-7.0
Okra	6.0-8.0	Rutabaga	6.5-7.5	Weigela	6.0-8.0
Onion	6.0-7.0	Rye	6.0-7.0	Wheat	6.0-7.0
Orange	5.0-7.0	Sage	6.0-8.0	Whitecedar	4.0-5.0
Oregon Holygrp.	6.0-8.0	Salsify	6.5-7.5	Willow	6.0-8.0
Oxalis	5.0-8.0	Salpiglossis	6.5-7.5	Wintergreen	5.0-6.0
Packysandra	5.0-8.0	Sandwort	4.5-5.5	Wisteria	6.0-8.0
Pansy	6.0-8.0	Saxifrage	6.0-8.0	Witch-hazel	6.0-7.0
Parsley	5.0-7.0	Scabiosa	6.5-7.5	Yew	5.5-7.0
Parsnip	6.0-8.0	Shadblow	6.0-7.0	Yucca	6.0-8.0
Partridgeberry	5.0-7.0	Shagbark Hick.	6.0-7.0	Zinnia	6.0-8.0

The correct pH is necessary for a plant to assimilate the nitrogen, phosphorus, potash, and other nutrients from the soil. In a soil with low pH (more acid), the availability of calcium, magnesium, nitrogen, and phosphorus decreases while manganese and aluminum may become more available and cause a toxic build up. In a soil with a high pH (alkaline—pH 8 or more), phosphates may be less available and plant growth may be inhibited because of lack of copper, zinc, and manganese.

In an alkaline soil, certain minerals such as iron, manganese, and copper may be fixed in chemical compounds. These minerals are then unavailable in solution for absorption by plant roots unless something like sulfur is added to increase soil acidity. *A laboratory soil test may show that plenty of the minerals is available in the soil; however, the plants may show signs of iron or manganese deficiency. This is a case where a professional soil test can mislead an untrained person. The minerals are present in the soil, but they are not available in a form which is usable by the plant.*

In acid soils, minerals such as manganese and aluminum can become so available that they can cause a build up in the soil which is then dangerous or toxic, to the plant. In either condition (extremely acid or alkaline), important trace elements, such as phosphorus, may be fixed in compounds which are unavailable to plants. These elements are generally available at between pH 5.5 and pH 7.0. The following paragraphs describe the effects of soil pH on some of the important nutrients and minerals in the soil.

3. Effects on Some Nutrients By pH

a. *Nitrogen.* Soil nitrogen is in a constant state of transformation by micro-organisms (nitrifying bacteria) when the soil temperature is above 50°F. Nitrification can take place between pH 5.5 to pH 8.0. Nitrifying bacteria, like plants, need an adequate supply of calcium and phosphate, and a proper balance of iron, copper, zinc, and magnesium. These nutrients are most available in a soil between pH 6.5 and 7.0. The process whereby micro-organisms decompose organic matter to release nutrients locked in dead plant and animal matter is also dependent upon a near neutral soil pH level.

b. *Phosphate.* The availability of phosphate to plants is reduced when the soil is below pH 6.0 because of the formation of relatively water-insoluble iron and aluminum phosphates. In a soil of around pH 6.5, the availability of iron and aluminum is reduced, allowing phosphorus to be made available to the plant. At values between pH 7.0 and pH 8.5, phosphate ions are again made unavailable to plants by formation of water insoluable phosphates of calcium and magnesium. Generally, phosphorus is most readily available between pH 6.5 and pH 6.8.

c. *Potassium.* The potassium in materials added to an acid soil (low pH) has great difficulty in replacing the hydrogen ions absorbed on the surface of soil particles. Since an acid soil can hold few potassium ions, these are free to move away from plant roots in water solution and be leached. When potassium-bearing materials are added to a soil which is nearly neutral (pH 6.5 to pH 7.0), a large percentage of the potassium ions get attached to soil particles through ion replacement. Thus, maintaining a proper pH makes more potassium available to your plants.

d. *Magnesium.* The same principle applies to the behavior of magnesium in the soil as described in the preceding paragraph for potassium. In humid regions with coarse-textured soil which has a tendency toward low (acid) pH soil magnesium deficiencies can develop because of the difficulty with which magnesium is held by the soil. As with potassium, magnesium is more readily available in soils with a near neutral pH.

e. *Sulfate.* Almost all of the inorganic sulfur in well-drained soils is in the form of sulfate ion in combination with magnesium, calcium, potassium, ammonium, and sodium. Sulfate ions possess a negative charge and are not readily absorbed by clay surfaces. This means that they can be easily leached from the soil. Leaching occurs most readily at pH 6.0 where there is a predominance of calcium, magnesium, potassium, and sodium ions present. Leaching losses are least where soils are acid, and appreciable amounts of exchangable iron and aluminum exist. Because of this, it is often necessary to use sulfur additives to grow plants with high sulfur requirements.

f. *Trace Elements.* The chemistry of trace elements such as iron, zinc, copper, and manganese varies greatly with changes in soil pH. In highly acid soils (pH 4.5 to pH 5.5), these trace elements are all highly water soluble and readily available to plants. In some cases, as previously noted, these trace elements are so readily available that they can reach concentrated proportions which are *toxic* to plants. To prevent toxicity and trace element imbalance, acid soils require lime to partially limit the availability of these trace elements to plants. However, as the soil is raised to pH 7.0 and higher, trace element availability is severely limited by the formation of water-soluble compounds such as oxides, hydroxides, carbonates, and bicarbonates. This is particularly true in the case of iron. To ensure that plants receive an adequate supply of the trace elements in nontoxic proportions, the soil should be maintained around pH 6.5.

Generally, then, at a soil value of pH 6.5 to pH 7.0, there will be sufficient calcium, magnesium, and potassium bound to the soil particles to maintain balanced plant nutrition, the trace elements and phosphate will be in a form that is available to plants, and nitrogen transformation will take place at a desirable rate. This, of course, presupposes that the soil contains adequate nutrients and minerals to replace those used by the plants or leached from the soil by rain or watering.

4. Testing Soil pH

Although the preceding probably makes one think that testing soil for pH requires a great deal of technical knowledge, it is actually very simple. Several methods are available from which one can choose.

a. The simplest test involves the use of neutral litmus paper, which is available at most drugstores. To use litmus paper, ensure that the soil is wet and press a strip of litmus paper into the soil. When the paper is moistened, it will change colors. If it becomes red or pink, the soil is on the acid side. If it turns blue, the soil is on the alkaline side. No color change means that the soil is neutral or pH 7.0. Various shades of red, pink, or blue indicate the degree of acidity or alkalinity: the redder, the more the soil is on the acid side; the bluer, the more the soil is on the alkaline side. Color charts are available for use with litmus paper so that one can readily determine the approximate pH from the litmus paper color.

b. More exacting are the soil test or pH kits normally available at most garden supply stores. To make the chemical type pH test, dry soil is mixed with the solution provided in the kit. This solution should be allowed to stand for a few minutes, and then be compared with a color chart to determine the pH. When using the method, care must be taken to use clean tools and containers which are free from any material that might contaminate the soil and affect the accuracy of the reading.

c. The fastest, simplest, and most accurate method of determining soil pH is to use a direct-reading soil tester or pH meter. This meter provides an accurate reading when its plates are pushed into the moist soil. The plates must be kept clean and wiped between each test with fine sandpaper to ensure good contact with the soil.

When testing soil pH, readings should be taken at different depths in the soil (i.e., top, 6 inches deep, and 10 inches deep). This will ensure that entire soil growing area is within the correct pH range.

Many factors can affect soil pH since soil consists of many different materials and is affected by the natural environment as well as the man-made environment. The basic soil itself may be one pH, any organic additive or fertilizer another pH, and the water something else. When all of these are combined in the soil, the pH should be between the extreme readings, depending on the proportional amount of materials. If the pH level is considerably to the acid or alkaline side, each material should be tested separately to determine which material is the cause.

Professional Laboratory

d. Most home gardens generally do well without a professional laboratory soil analysis. The home gardener will usually determine what grows well in his soil by observation or trial and error, and what additives are necessary for maximum production. If serious problems do occur, a professional analysis, conducted by the county agricultural agency or an independent test laboratory, can pin point the problem. Some of the major problems in obtaining abundant plant growth are lack of certain nutrients, poor or irregular watering practices, insects or diseases, salinity, or toxic substances in the soil.

5. Controlling pH

Some treatment of the soil is generally necessary to grow fruits, vegetables, and grains in the United States since most soil is either acid or alkaline. Cultivated soils are even more acid than normal because the natural alkaline controlling agent (lime) is easily leached from cultivated soils. Soils which have been mulched or covered with sod generally maintain their alkalinity for a longer period of time.

While decaying organic matter, commonly known to gardeners as *compost,* can help to correct soil acidity, it works very slowly. In cases of extreme acidity, humus may not contain enough alkalinity to counteract the acid condition. Additionally, certain organic matter such as peat moss, pine needles, cottonseed meal, and oak leaves are highly acid, and could contribute to the acid condition. Test soil pH sometime after each addition of compost.

Gypsum (calcium sulfate) is a neutral salt, and although it is recommended by some gardening books, it does *not* change soil pH.

a. *Controlling Acidity.* An acid condition in the soil is easily corrected by use of calcium carbonate (limestone flour). *Limestone,* or calcium carbonate, is the material used with chicken feed for eggshell development. Optionally, crushed oyster shell or other crushed calciferous shell may be used. If the pH is below the correct level (on the acid side) for the desired plant growth, it can easily be corrected by sprinkling limestone over the soil and then watering it into the soil. Five pounds of limestone sprinkled over a 10-foot by 10-foot area (100 sq. ft.) of loamy soil and watered into the soil will raise the soil pH by about one point for a period of approximately three years. Table 13.4 lists the approximate amount of limestone to use for loamy, sandy, or clayey soils. The quantities can be multiplied by the area to obtain the correct amount of limestone to use for larger or smaller areas; that is, for 25-square feet, use 1¼ pounds.

Wood ashes can be used to correct an overly acid soil condition in the same manner as calcium carbonate. However, large quantities of wood ashes should not be applied to the soil unless the soil definitely has a low pH since this could make the soil too alkaline for many plants. Wood ashes should be used at approximately double the rate suggested for limestone in Table 13.4 (see next page).

Decomposed (composted) *cardboard* may also be used to help neutralize acidity; it is generally neutral, or slightly alkaline.

In many cases, a thorough *watering* alone will bring the

pH up to the required level. Most city water is on the alkaline side (above pH 7.0), and pond water is sometimes above pH 10.0.

TABLE 13.4—Soil treatment for adjusting pH with lime, alum or sulfate.

To change pH from	to	Material to add	For loamy soil add		For sandy soil add		For clayey soil add	
			per sq. ft. (teaspoons)	per 100 sq. ft. (lbs.)	per sq. ft. (teaspoons)	per 100 sq. ft. (lbs.)	per sq. ft. (teaspoons)	per 100 sq. ft. (lbs.)
4	5	lime	3	5	2½	4	3½	6
5	6	lime	3	5	2½	4	3½	6
6	7	lime	3	5	2½	4	3½	6
5	4	alum or sulfate	1	4	¾	3¼	1¼	4¾
6	5	alum or sulfate	1	4	¾	3¼	1¼	4¾
7	6	alum or sulfate	1	4	¾	3¼	1¼	4¾

b. Controlling Alkalinity

The limited rainfall in the more arid areas of the United States has resulted in an accumulation of calcium carbonate in the soil. The lower the rainfall, the nearer the surface that this accumulation will occur. As a result, these soils may have alkaline reactions. In some areas, sodium has accumulated to various degrees. High sodium-content soils may have values above pH 8.5.

In many areas, the subsoil may contain a limestone layer. This layer can cause a process known as percolation whereby lime will continually "percolate" to the surface or into the topsoil layer. In this case, it is generally necessary to treat the soil each year for an overly alkaline condition.

Alkalinity can be controlled chemically by the addition of sulfur or aluminum sulfate, or organically by the addition of peat moss, rotted sawdust, pine needles, oak leaves, and cottonseed meal. The correct amount of chemical additive necessary for controlling alkalinity is listed in Table 13-4. Either sulfur (preferred) or aluminum sulfate should be added to the soil in the same manner as described in the preceding paragraphs for limestone.

Sulfur is the most widely used soil amendment for lowering pH. Sulfur, incidentally, is important for the development of plant and animal tissue; the oxidizing reaction produces energy which is used by the micro-organisms. When sulfur is added to the soil, micro-organisms *Thiobacillus oxidans*) convert the sulfur to sulfuric acid. The sulfuric acid produced in this reaction is ionized, and releases hydrogen ions which cause a reduction of soil pH.

One of the fastest remedies for an overalkaline condition is to mix *dry peat moss* into the soil. Most peat moss has a value of between pH 3 and pH 5 in its dry state. Another method is to use decomposed, or composted, *newsprint,* which is only slightly acid. When using organic matter to control alkalinity, periodic tests must be made to determine when the correct pH level is achieved.

The addition of sulfur or limestone to a soil may take several weeks to have full effect. Therefore, if pH testing shows either is necessary, this should be one of the first steps in preparing the soil before planting time. As the pH values of a soil affect very greatly our plant growth, it is important to control these levels to whatever is ideal for your crops.

★ ★ ★

CHAPTER 14

ESSENTIAL NUTRIENTS FOR PLANT GROWTH

A. SIXTEEN ELEMENTS

It is known that 16 elements are required in the soil to support higher (non-microscopic) plant life. These elements give soil its fertility. Natural soil fertility means that a soil has a balance, or near balance, of the 16 nutrient elements as well as the plant supporting characteristics discussed in the preceding chapter.

The 16 essential nutrients for plant growth which come from three primary sources: air, water and soil, are:

1. nitrogen (N)
2. phosphorus (P)
3. potassium (K)
4. sulfur (S)
5. magnesium (Mg)
6. calcium (Ca)
7. boron (B)
8. iron (Fe)
9. zinc (Zn)
10. manganese (Mn)
11. copper (Cu)
12. molybdenum (Mo)
13. chlorine (Cl)
14. carbon (C)
15. oxygen (O)
16. hydrogen (H)

Carbon, hydrogen, oxygen and nitrogen are classified as non-metallic elements. H, O and N are gases and carbon can combine with other elements to form a gas. Carbon is supplied to the plant from the atmosphere as carbon dioxide; oxygen and hydrogen are provided from the air and from water; and nitrogen is provided through fixation of nitrogen from the air by the soil, and from decomposing organic matter. The other nutrients are normally present in the soil and come primarily from the surface rock particles which were broken down to make the soil of that area.

Plants will grow well if enough of all 16 nutrients are available to the roots in water or soil solution. However, if the soil is deficient in one or more nutrients, the overall balance is upset, and the plants will show signs of nutrient deficiency. Plants, in many ways, are like human beings. They have a very sensitive metabolism which must receive the proper balance of nutrients and vitamins for healthy growth.

The elements nitrogen (N), phosphorus (P) and potassium (K) are used by plants in greater quantities than any of the other elements. Therefore, they are known as the primary elements.

Calcium (Ca), sulfur (S) and magnesium (Mg) are known as secondary elements. A good supply of these nutrients is generally available within the soil.

The remaining seven nutrients: boron, iron, zinc, manganese, copper, molybdenum, and chlorine, are known as trace elements or micronutrients. These nutrients are just as important to plant growth as the others, but are required in only minute amounts. Most soils contain an adequate supply of these nutrients except in situations of extreme soil acidity or alkalinity or for certain deficiencies in large areas for specific crops. A good continuous supply of organic matter releases most of these elements into the soil.

The role of the earthworm in the release of nutrients or in making nutrients available to plants is not fully known or understood. However, as explained in previous chapters, the earthworm has a very important role in nutrient availability.

In most cases, the methods by which plants obtain the 13 mineral or metallic nutrients and nitrogen is quite complex as is the functioning of those nutrients within the plant. This chapter discusses each of these nutrients in depth to provide a better understanding of how the nutrients function in the soil and in the plant. The functioning of carbon, hydrogen and oxygen is interrelated to the reactions of the mineral elements in both the soil and the plant. Table 14-1 provides a quick reference summary of plant nutrients, their functions in the soil, signs of deficiency, signs of excess and the correct soil pH level required to obtain maximum nutrient availability.

B. SUMMARY OF ESSENTIAL ELEMENTS FOR PLANT GROWTH

1. NITROGEN (N)
Form Absorbed: *Nitrate (NO_3^-)*, Ammonium (NH_4^+)
Best pH: pH 6.5 to pH 7.0
Function:
- —plants use N during early plant growth,
- —young plants may take up large amounts and store them,

- generally concentrated in rapidly growing areas such as buds, shoots, newly open leaves,
- regulates plant's ability to make protein,
- causes deep green leaf color by increasing chlorophyll,
- encourages above ground vegetative growth necessary for fruit and seed development,
- increases plumpness and protein content of grain crops,
- produces succulence in crops—grass, lettuce, radishes, etc.
- NO_3^- form is highly mobile in plants,
- highly water soluble, easily leached from soil.

Signs of Deficiency:
- paling and yellowing of leaves (chlorosis), including veins, leaves may fall off, older leaves are affected first,
- stunted plant growth, smaller leaves, fewer flowers and smaller fruit, delay in bud opening and development of flowers and leaves.

Signs of Excess:
- plants grow too fast, become weak and spindly, flower late in season,
- resulting overleafy, thin-stemmed plants more susceptible to disease, drought and cold.

2. PHOSPHORUS (P)

Form Absorbed: Dihydrogen Phosphate ($H_2PO_4^-$), Dicalcium Phosphate, ($CaHOP_4$)

Best pH: pH 6.5 to pH 6.8

Function:
- continually taken up during life of plant at varying rates, slowly slackening off as plant ages,
- an active element of living material (protoplasm) in plant cells, essential for many life processes,
- provides vehicle by which energy, released by burning of sugars, is transferred within plants,
- encourages root growth, especially lateral and fibrous roots,
- flowering and fruiting rates are dependent on P,
- hastens crop maturity,
- increases yield ratio of grain to straw,
- improves crop quality and disease resistance,
- mobile in plant.

Signs of Deficiency:
—many symptoms similar to N deficiency,
—growth of entire plant stunted, slow root development, poor root system, delayed maturity, lack of or poor quality fruit and seed development,
—extreme deficiency indicated by dull green leaves with purple tinges.
Signs of Excess:
—little danger of excess available of P in the soil.

★ ★ ★

3. POTASSIUM (K)
Form Absorbed: K^+, often expressed as potash (K_2O)
Best pH: pH 6.5 to pH 7.0
Function:
—essential to manufacture and movement of sugars and starches within plants and to normal growth through division of plant cell,
—may hasten maturity and seed production,
—aids in root development,
—increased resistance to disease,
—delays maturity, counteracting effect of P,
—needed for tuber, chlorophyll and starch formation.
Signs of Deficiency:
—growth slower than normal,
—more susceptible to disease,
—leaves with mottled yellow tips and edges,
—older leaves may look scorched around edges.

★ ★ ★

4. SULFUR (S)
Form Absorbed: $SO_4^=$
Best pH: pH 6.3 to pH 6.8
Function:
—increases root system of plant and nodulation of legumes,
—promotes chlorophyll formation,
—required in synthesis of amino acids,
—improves physical condition of the soil,
—lowers pH of naturally alkaline soils,
—increases availability of other plant nutrients such as phosphorus, iron and manganese,
—immobile in the soil.

Signs of Deficiency:
—similar to that of nitrogen,
—plants are small, spindly, have short slender stocks,
—growth rate usually retarded and maturity delays,
—young leaves often light green to yellowish in color with even lighter-colored veins,
—retardation of nodule formation on legumes.

★ ★ ★

5. MAGNESIUM (Mg)
Form Absorbed: Mg^{++}
Best pH: pH 6.5 to pH 7.0
Function:
—only mineral element constituent of chlorophyll molecule, thus necessary for chlorophyll formation,
—increases nodulation of legumes,
—necessary to activate enzyme system and affects oil content of plant,
—mobile in plant.
Signs of Deficiency:
—loss of green color between veins followed by yellowing of leaves starting at leaf tips or margins and progressing inward,
—abnormally thin leaves, leaves brittle and curve upward,
—severe deficiencies cause leaves to become reddish purple color on tips and along edges.

★ ★ ★

6. CALCIUM (Ca)
Form Absorbed: Ca^{++}
Best pH: pH 6.3 to pH 6.8
Function:
—associated with activating the enzyme system,
—encourages root and leaf growth,
—increases nodulation of legumes,
—controls soil pH.
Signs of Deficiency:
—gelatinous, sticky substance occurs at tips of new leaves, causing them to stick together.

★ ★ ★

8. IRON (Fe)
Form Absorbed: Fe^{+++}, Fe^{++}
Best pH: pH 6.3 to pH 6.8
Function:
—related to chlorophyll development,
—functions specifically in activating plant enzyme systems,
—immobile in plant.
Signs of Deficiency:
—deficiency causes lack of chlorophyll development,
—interveinal yellowing of new leaves, progressing to older leaves.

★ ★ ★

9. ZINC (Zn)
Form Absorbed: Zn^{+++}
Best pH: pH 6.3 to pH 6.8
Function:
—metal activator of enzyme system for growth in plants,
—functions in protein metabolism,
—toxic in excess of trace amounts.
Signs of Deficiency:
—reduced plant vigor, slower growth, delayed maturity,
—reduced yields, lower quality fruits.

★ ★ ★

10. MANGANESE (Mn)
Form Absorbed: Mn^{++}
Best pH: pH 6.3 to 6.8
Function:
—activates enzyme systems in plants,
—assists in photosynthesis.
Signs of Deficiency:
—defoliation, loss of vigor, lower yields,
—interveinal chlorsis.
Signs of Excess:
—stunting of plants and yellowing of leaves.
—occurs mostly in acid soils.

★ ★ ★

11. **COPPER (Cu)**
    Form Absorbed: Cu^{++}
    Best pH: pH 6.3 to pH 6.8
    Function:
       —vital part of several enzyme systems,
       —regulates some life processes,
       —functions in chlorophyll formation.
    Signs of Deficiency:
       —lower yield and lower quality,
       —stunted growth, yellowing of plants, deficiency occurs mainly in soils high in organic matter which are alkaline soils with a high pH.

★ ★ ★

12. **MOLYBDENUM (Mo)**
    Form Absorbed: Mo_4^{++}
    Best pH: pH 6.3 to pH 6.8
    Function:
       —activates enzyme systems,
       —toxic in excess of trace amounts.
    Signs of Deficiency:
       —interveinal chlorsis,
       —legumes turn pale and yellow.

★ ★ ★

The other essential nutrients are included with the ones summarized in the following expanded explanations. These include:

    7. boron (B)
    13. chlorine (Cl)
    14. carbon (C)
    15. oxygen (O)
    16. hydrogen (H).

C. NUTRIENTS IN DETAIL

1. NITROGEN

Nitrogen is a basic building block of all living matter. Of all the 16 essential nutrients for plant growth, it is probably the most important. Since nitrogen is used by plants in relatively large amounts it is called a primary nutrient. Also, nitrogen is a basic part of the living protoplasm of all cells—LIFE is impossible without nitrogen. Webster's New

Collegiate Dictionary defines nitrogen as "a colorless gaseous element, tasteless and odorless, constituting about four-fifths (78.03%) of our atmosphere by volume, and a constituent of all living tissues."

Nitrogen is a unique element as it exists as an inert gas in large quantities in the air. It is estimated that approximately 35,000 tons of nitrogen exist above each acre of land. This inexhaustible supply remains fairly constant with nitrogen being returned to the atmosphere at approximately the same rate that it is removed. It would seem with all this nitrogen that plant growth would never suffer from a shortage. However, nitrogen is probably the most common of all nutrient deficiencies. The inert abundant nitrogen gas of the atmosphere, N_2, cannot be used directly by the plants. It must be chemically fixed or combined with other elements in the soil before higher plant life can use it for growth.

Nitrogen in the soil is unique in that it does not occur in mineral form. Special soil micro-organisms must combine gaseous nitrogen from the air with other elements in a process called "nitrogen fixation" before it is available to the plants. Some of these nitrogen fixing micro-organisms exist in nodules on plant roots, primarily the roots of legumes such as peas. Other micro-organisms which fix nitrogen live freely in the soil.

While some micro-organisms "fix" nitrogen, others which live on decaying organic matter release nitrogen in a usable form for plants as they break down the organic matter. Nitrogen which enters the soil from dead plants and animal material undergoes several changes before it can be absorbed by plant roots and used within the plant. First, one group of micro-organisms digest the raw nitrogen and convert it to a new form, ammonia (NH_4^+). Another group of soil micro-organisms then changes the ammonia to the nitrate (NO_2^-) form of nitrogen, and then still another group converts the nitrite to the nitrate (NO_3^-) form of nitrogen which is usable in plants.

The supply of usable nitrogen in the soil controls a plant's ability to make proteins which are necessary for the growth of each plant cell. This promotes rapid growth of stems and leaves and provides plants with a deep green color. An adequate nitrogen supply is especially important for growing leaf groups such as lettuce or cabbage. Nitrogen, however, is leached through the soil beyond the reach of plant roots very rapidly by rain or constant watering. Thus a source of nitrogen must be added often and in large amounts to the soil.

a. Nitrogen Fixation

Nitrogen fixation is a biological conversion of atmospheric or elemental nitrogen (N_2) to a combined form of nitrogen with other elements by microbial activity. Only a few micro-organisms, some bacteria and certain algae, possess the ability to fix nitrogen into a form usable by the biological systems of plants and micro-organisms.

Elemental nitrogen from the air is assimilated by micro-organisms as building material for cells within the organism. Nitrogen is converted to ammonium or hydroxylamine, then to amino acids and finally to cell protein. When the cell dies, the nitrogen component in the protein is again converted to ammonium which can be used by other micro-organisms or absorbed by plants.

Ammonium or amino acids are formed from atmospheric nitrogen more rapidly than they can be used by the cells under certain conditions. In that case, some soluble nitrogen compounds are excreted from the cells and become available for use by other micro-organisms or by plants.

Four general groups of micro-organisms involved in the fixation of nitrogen in the soil are:

  a. Legume-nodule bacteria or *Rhizobium*
  b. Free-living or *Azotobacter* — aerobic bacteria which are living or active only in the presence of oxygen.
  c. Free-living or *Clostridium* — anaerobic bacteria which live or are active in the absence of any free oxygen.
  d. Free-living aerobic green plants or *Algae*.

All of these, except possibly the *Azotobacter*, contribute substantially to the nitrogen availability of the soil.

b. Nitrogen Absorption By Plants

Most plants tend to absorb the greatest part of their nitrogen requirements during the early stages of their growth. Many plants absorb up to 2/3 of their total nitrogen requirements during the first 60 days of their life cycle even though some of this nitrogen may be stored for later use. The nitrogen is generally concentrated in the rapidly growing areas of the plants such as buds, shoot tips and new leaves.

C. Nitrogen Cycle and Mobility in Soil

Nitrogen goes through a definite pattern commonly known as the "nitrogen cycle". Nitrogen which is fixed or chemically combined becomes part of the organic matter in the soil over a period of time. As this supply of material decomposes, some of the nitrogen is released for plant growth and some is released into the atmosphere, thus replenishing what had been removed.

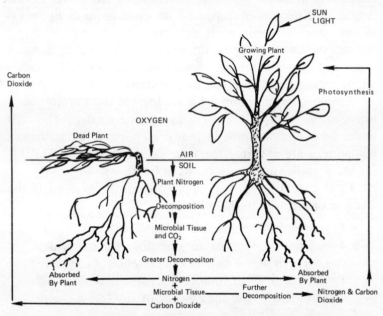

FIGURE 14.1—Nitrogen cycle and mobility in soil

In some soils, it has been estimated that approximately 1,000 to 3,000 pounds of nitrogen per acre is contained in the top six inches of soil and of this only a little more than 1%, or 10 to 30 pounds per acre, is in a form which is immediately usable by plants. The remainder of this nitrogen supply is slowly converted to a usable form over a period of time.

Nitrogen is either a mobile or immobile nutrient in the soil, depending on whether it is in the ammonic (NH_4^+) form or the nitrate (NO_3^-) form. The + or − signs on the equation indicate the electrical charge of the particle. This electrically charged particle is known as an "ion". Nitrogen combined with hydrogen forms ammonium (NH_4^+) ions which have a positive electrical charge while nitrogen combined with oxygen forms (NO_3^-) ions which have a negative charge. The electrical charge is what determines the mobility of the nitrogen carrier in the soil. Since soil particles or soil *colloids*, generally contain a predominantly negative electrical charge and since unlike charges attract, the positively charged ammonium ion is attracted to the soil particle and is held or bound to that particle. This means that the ammonium ion is immobile and cannot move through the soil solution. Nitrate (NO_3^-) is mobile and can move through soil solution around the soil particles. Ammonium is thus stored in soil but not directly usable by the plants while highly mobile nitrate moves readily and can be leached from the soil.

Plant roots can absorb either nitrate or ammonium, but it has been estimated that 90% of the absorbed nitrogen nutrient is in the nitrate form. As the immobile ammonium ion is only available to the root at the point that the root touches the ammonium-bearing soil particles, little NH_4^+ can be absorbed. However, plants absorb large quantities of water or soil solution and thus nitrate is taken in by the root system.

But later in the growing season, most of the available nitrogen is in the nitrate form because the ammonium form is converted to the nitrate form by micro-organisms in the soil. When so converted, it becomes a mobile nutrient. This conversion occurs under soil conditions that also permit rapid plant growth.

The movement of the nitrate form of nitrogen to the root surface with soil water brings up an important factor necessary for all plant growth. Nitrogen and *all* the nutrients must be in a moist soil or solution for assimilation by plants. For this reason, the downward movement of the nitrate form of nitrogen is highly important during dry weather.

d. Nitrification

Nitrification is the process whereby the immobile and generally unusable ammonium ions in the soil are converted to mobile and usable nitrate ions. The major steps from ammonium to nitrite to nitrate are accomplished by *Nitrosomanas* and *Nitrobacter* bacteria.

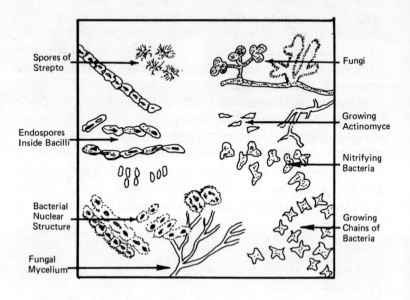

FIGURE 14.2—Bacteria
Several types of micro-organisms that may appear in your compost pile

e. Denitrification

A reverse reaction to nitrification occurs under certain soil conditions whenever air is excluded from the overall system. The process, known as denitrification, is one which reduces nitrates to the elemental nitrogen gas. Denitrification results in loss of nitrogen back to the atmosphere and away from the soil as a nutrient.

Certain micro-organisms which use oxygen in their metabolism take their oxygen from the nitrate ions and release the nitrogen gas. Denitrification happens in water-logged, very compacted soils or ones which contain large amounts of carbon-type materials such as straw.

f. Mobility of Nitrogen in Plants

Young plants absorb and store nitrogen which is then transported from the roots to the leaves and stems, and then into the developing grain or food-storage areas. Once nitrogen is absorbed in some form into the plant system it is not subject to loss by leaching or denitrification. Early application of nitrogen to vegetables or grains will enhance tillering and vegetative growth while applications at or before maturity can greatly increase the protein content.

g. Nitrogen Compounds in Plants

Once a nitrogen compound is absorbed in the plant, it is converted to the amino form, NH_2^- by a process called nitrate reduction. Additional chemical changes occur as the products of photosynthesis are combined with amino groups to form amino acids—the building blocks of proteins. Some nitrogen compounds which are essential for plant growth and production are:

i. Amino Acids

These organic acids contain one or more amino groups which are the building blocks for protein, a constituent of every living cell. Protein is a major component of protoplasm which occupies the interior part of living cells. The formation and growth of plant cells requires a continual supply of protein.

ii. Nucleic Acids

DNA and RNA are two well-known nucleic acids involved in cell reproduction. Plant growth requires two basic steps, cell division and cell enlargement. For cell division, these nucleic acids must reproduce themselves. This process depends upon a continued supply of nitrogen.

iii. Chlorophyll

Chlorophyll is the green pigment responsible for the process of photosynthesis. Photosynthesis uses the sun's energy (light) to combine water and carbon dioxide to form essential carbohydrates and sugars for plant growth.

h. Nitrogen Deficiency in Plants

Nitrogen deficiency can cause these four results:

a. Stunting of growth will result since nitrogen is necessary for cell division and enlargement.

b. The yellowing of tissue is called chlorosis. Chlorophyll causes the green color of plant tissues. When nitrogen is deficient, the production of chlorophyll decreases or stops. The tissue becomes yellow due the presence of yellow pigments within the cell. These yellow pigment compounds are present in green cells also, but they are obscured by the green chlorophyll molecules.

c. Nitrogen compounds in older tissues break down first, and the nitrogen then moves to younger tissues. Thus, the chlorosis will first occur in the tips of older leaves. As the deficiency becomes more acute, the entire plant may become chlorotic. The bottom leaves often drop off and yields are frequently reduced, especially if the deficiency occurs early in the plant life cycle.

d. Since nitrogen is a constituent of protein, a deficiency can affect vegetative protein content. However, transportation of absorbed nitrogen from leaves and stems to the grain or fruit will often compensate for some of the lack of nitrogen in the soil. This maintains a reasonable protein level. A nitrogen deficiency may be caused by an incorrect carbon-to-nitrogen ratio.

i. Excess Nitrogen in Plants

Normal amounts of nitrogen promote healthy growth and slightly hasten the maturity of most plants. However, excessive amounts of nitrogen can cause excessive vegetative growth, delaying the maturity process. Too much nitrogen causes plants to grow too fast, become weak and spindly. Also the plants and fruit flower late in the season. The resulting overleafy, thin-stemmed plants may be more susceptible to disease, drought and cold. This condition may also be caused by an excessive carbon-to-nitrogen ratio.

j. Carbon/Nitrogen Ratio

The transformation of nitrogen in the soil is controlled almost exclusively by the micro-organisms in the soil. The decomposition of plant residues depends upon the amount of nitrogen and carbon, as well as the source of the carbon. Decaying organic matter in the soil includes carbon and nitrogen and other elements. The soil micro-organisms feed upon this organic matter as they decompose it, using the nitrogen as a building material and the carbon (in the form of

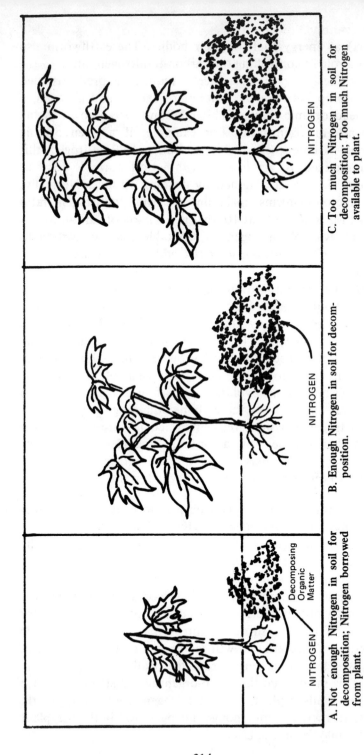

FIGURE 14.3—Effects of carbon/nitrogen ratio on plants

sugars) as energy fuel for their bodies. The earthworm also plays a role in controlling the carbon-to-nitrogen ratio.

Synthesis by microbial cells requires a predetermined relationship between the amount of carbon and nitrogen in the dead organic matter. This relationship is known as the carbon-to-nitrogen ratio. For example, if a given organic material is composed of 30% carbon and 2% nitrogen; this ratio reduced to its lowest arithmetical form is 15:1 or simply 15. This is obtained by dividing 30 by 2. Normally, the micro-organisms maintain a carbon-to-nitrogen ratio *within the soil* of about 10; that is, 10 parts of carbon to one part of available nitrogen. A workable ratio of carbon to nitrogen *in any organic material* added to the soil is 30:1 or 30.

During decomposition of organic material, nitrogen is liberated only when the supply of available nitrogen is greater than that required by the micro-organisms. This liberated nitrogen may be recycled by the micro-organisms until the carbon content decreases to a level where the carbon-to-nitrogen ratio is below 30. This narrowing of the ratio is accomplished by the evolution of carbon dioxide during the biological oxidation of the carbon-containing plant component. Nitrogen is continuously recycled in various combined forms. This is necessarily a continued recycled process since most plants cannot assimilate mineral nitrogen unless the carbon-to-nitrogen is near 20:1 or even lower.

The normal carbon-to-nitrogen ratio in the soil becomes unbalanced when an organic material which is high in carbon and low in nitrogen, such as straw, is added to the soil. The high carbon content provides additional energy material, stimulating the micro-organisms in the soil and causing them to increase in number. As the micro-organisms burn up the carbon to obtain energy, they need additional nitrogen for growth. If there is sufficient nitrogen in the added organic material to meet the growth requirements, the micro-organisms will borrow nitrogen from the available supply in the soil. This means that the micro-organisms must compete with the surrounding plants for the nitrogen needed for growth. Thus, if there is not enough nitrogen for both, the plant growth may be stunted.

This is only temporary. The nitrogen may be locked up in the bodies of the micro-organisms when needed most by plants, but when the ratio can no longer support the increased number of micro-organisms, they start to die. The dead micro-organisms then decay and return nitrogen to the soil.

Conversely, when the nitrogen content of added organic matter is too high, the resultant can be an excess of nitrogen in the plant. The decomposition of any organic material with between 1.25 and 1.5% nitrogen on a dry-weight basis is essentially self-sustaining. That is, the carbon-to-nitrogen ratio is 30:1 and nitrogen is not borrowed from the soil by the decomposition process.

Plants, such as the legume (bean, pea) family, are the highest in nitrogen content and have a high carbon-to-nitrogen ratio. Peat moss, leaves, leaf mold, manure and compost generally contain sufficient nitrogen for self-sustaining decomposition. But straw, sawdust (particularly pine, fir and redwood), grain stubble and sorghum stalks have a high carbon-to-nitrogen ratio. One of the following methods should be used to ensure a correct carbon-to-nitrogen ratio when adding organic matter to the soil.

a. When adding an organic material with a high carbon-to-nitrogen content, a natural fertilizer, too. This is highly important when fresh organic matter is added during the growing season. The added nitrogen keeps the increased microbial activity from temporarily removing nitrogen from the soil, thereby preventing retardation of plant growth.

b. Any fresh, high-carbon/low-nitrogen content organic matter should be added to the soil several months before planting to allow nitrogen to be returned to the soil through death and decay of the micro-organisms. Organic matter added in autumn will begin to decompose and the nitrogen will be locked in the microbial cells during the winter. Thus the nitrogen will not leach away during heavy winter rains. Then the nitrogen will be slowly released into the soil in the spring when the plants need it the most.

c. Composting discussed in Chapter 15 is the best way to break down materials without disrupting the carbon-to-nitrogen ratio of the soil because the initial decomposition process occurs outside the soil. Composting allows control of the nitrogen, air, water and heat during microbial decomposition. The finished or partially finished compost can then be placed on or in the soil without danger of disrupting the carbon-to-nitrogen ratio.

2. PHOSPHORUS

Phosphorus is the second of the three primary nutrient elements that are essential for plant growth. Except for nitrogen, unsatisfactory plant growth is usually due to the lack of available phosphorus in the soil rather than any of the remaining nutrient elements. Phosphorus is an active constituent of the protoplasm or living material in plant cells. It is necessary for the production of energy-giving sugars within plants, functions in the metabolic processes of the plant, and is the carrier whereby energy, released by the burning of sugars, is transferred within the plant.

Phosphorus in the soil encourages plant root growth, especially the lateral and fibrous roots. It hastens crop maturity and increases the yield ratio of grain to straw as well as total grain yield. Phosphorus also improves the quality of plants and increases disease resistance.

While soil may contain large quantities of phosphorus, it is generally in a form which is not readily available to plants or the form may be released too slowly. Phosphorus absorption by plants, like nitrogen absorption, is limited by the solubility of the phosphate compounds in the soil. Although the total quantity of phosphorus in a soil may range from 200 pounds per acre to 2,000 pounds per acre, only a small part of this, 0 to 200 pounds per acre or 0.1 parts per million to 1 part per million (ppm), is actually available to the plant.

Phosphate in the soil is in direct contrast to nitrogen. The basic chemical forms of phosphorus are generally more complex than those of nitrogen and when added to the soil, even in a usable form, react rapidly with other elements to form several new compounds which are less soluble. Additonally, phosphate is immobile in the soil as it is generally in an insoluble compound and moves very little in the soil solution.

a. Phosphate Fixation

The conversion of phosphates to other compounds is known as phosphate fixation and is directly related to soil pH. Phosphate fixation or unavailability is higher in acid soils than in neutral or alkaline soils. Phosphorus is more readily available to plants at between pH 6.5 and pH 6.8. Another term for fixation is *reversion*.

As long as the phosphorus remains in the soil, it has a nutrient potential which may be realized as the changing soil environment stimulates other chemical reactions. Microbial activity can also play an important part in the availability of phosphorus as a plant nutrient.

b. Sources of Phosphorus

There are two basic organic materials which may be added to the soil as a phosphorus fertilizer, *rock phosphate* and *bone meal*. Also, there are two types of chemical fertilizers which are manufactured using basic rock phosphate: superphosphate and treble superphosphate. Both of the chemical fertilizers are treated with organic compounds—sulfuric acid to create *superphosphate* and phosphoric acid to create *treble superphosphate*—to create the solubility of the rock phosphate. From an ecological point of view, use of either may add too much sulfur or phosphorus to the soil and cause an unnatural reaction in microbial activity. However, limited application of either, preferrably the treble superphosphate, may be beneficial in adding the immediate phosphorus required by plants to the soil.

Although phosphate is an excellent natural source of phosphorus, it varies in composition depending on its source. The basic component of rock phosphate is approximately 60% (25 to 30% P_2O_5) calcium phosphate or bone phosphate. It also contains calcium carbonate, calcium fluoride, alumina, manganese, iron oxide, iron sulfide, copper and other minerals, many of which are essential for plant growth. The best type of rock phosphate for soil application is a finely ground one. The organic acids and carbon dioxide produced by plant roots, as well as the microbial activity in the soil, can make the minerals available as nutrients at a faster rate from small particles than from larger particles.

Some researchers speculate that the natural chelation process in the soil contributes significantly to the phosphorus which is available to plants. Also, earthworms are thought to contribute to making phosphorus available to plants.

The chemical term P_2O_5 stands for phosphorus pentoxide or phosphoric acid. This term is used in analyzing all organic and inorganic sources of phosphorus to state the amount of *available nutrient* content. The P_2O_5 content means that if the phosphate is converted to the acid form, it would have that content of phosphoric acid. The percentage of P_2O_5 multiplied by 0.44 equals the percentage of elemental phosphorus (P) in the product.

There are two principal forms of expressing phosphate solubility in the United States—one which is water soluble and one which is soluble in neutral ammonium citrate or citrate soluble. The sum of water-soluble and citrate soluble phosphorus constitutes available phosphoric acid (P_2O_5) in a given product. Chemically, these are defined as:

a. water soluble phosphate—$Ca(H_2PO_4)_2^-$
monocalcium phosphate

and b. citrate soluble phosphate—$Ca_2HPO_4^-$
dicalcium phosphate

c. Available Phosphate

At any one time, only a small part of the total phosphorus needed for absorption by the plants is available in soil solution. Phosphorus must be continuously released into the soil solution for plant growth. The rate of phosphorus release into the solution is dependent upon several factors, including soil pH and temperature.

Maximum phosphate availability occurs at a soil value of pH 6.5. Even at this pH value, only a small fraction of the added phosphate remains in a water-soluble form. In neutral to alkaline soils (pH 7 and above), the presence of lime (calcium carbonate) causes a precipitation (dropping out) of phosphorus. In acid soils (pH below 7), aluminum and iron react with phosphorus to form highly insoluble compounds. Also, different chemical forms of phosphorus vary greatly in solubility.

d. Soil Temperature

Soil temperature affects the solubility of chemical compounds. Phosphorus is more available in warm soils than it is in cold soils. Often winter plants will not respond to phosphate even though the phosphorus soil level is relatively high. Since roots also grow faster in warm soils, they inhabit more soil volume from which to absorb nutrients.

e. Mobility in Soil

Phosphate should be placed into the root zone for efficient use by the plant, as the low solubility of phosphorus makes it immobile. How the soil stops, fixes or ties-up phosphate movement in several ways has been discussed. Often this fixation makes the phosphate unavailable to the plants. In *acid* soils, the aluminum phosphate is usually predominant, but iron phosphate is also prominent. Under *alkaline* conditions, the phosphate tie-up is caused primarily by calcium reacting with the phosphates. Phosphate fixation increases as the soil becomes more and more acid or alkaline. Additionally, phosphate fixation will be in heavier clay soils than in lighter sand soils at any given pH. Phosphorus moves in soils almost exclusively in *organic combination*. Soil micro-organisms were found to be the major factor controlling phosphorus redistribution in the soil. The earthworm also plays an important role in the movement and the availability of phosphorus in the soil. However, the total amount of phosphorus moved is still small when compared to nitrogen.

The *organic method* is very helpful in moving the little phosphorus that can be made available at any one time. A highly specialized relationship exists between most plant species and micro-organisms. In some, a nodule is formed on the plant roots or leaves and the organisms in the nodules fix nitrogen for the plant, while the plant furnishes carbon and nutrients for the microbial components. Both members benefit and their association is referred to as a *symbiotic relationship*—between the plant and the fungi.

This relationship enables plants to obtain less readily available plant nutrients from the soil minerals and humus, especially *phosphorus*. Soil microbes have bodies rich in phosphorus which is returned to the soil when they die. Controlled studies have shown that the uptake of phosphorus is increased in the presence of fungus. Since there are more fungi in earthworm castings than in the surrounding soil, this will help explain increased growth of plants with added earthworms to the soil. The earthworm, as it burrows through the soil, contributes to the mobility and solubility of immobile and insoluble nutrients. Thus, there are several ways in which immobile and insoluble forms of phosphorus and other nutrients of the same characteristics are made available to plants.

f. Phosphorus Absorption by Plants

Plants have a continuous need for phosphorus. In annual and biennnial plants, the nutrient is absorbed at a rapid rate during the intensely active growth period. As the plants approach maturity, absorption rate slows down, but very seldom levels off or stops until the plant reaches maturity and dies. Phosphorus is continuously absorbed at varying rates over the entire year by evergreen crops such as citrus.

g. Phosphorus Compounds in Plants

Phosphate ions, unlike nitrate and sulfur ions, do not undergo a chemical change in the plant tissue. The phosphate ions from the soil solution are first attached to the surface of the root and then passed through the root cell wall by means of a complex "carrier" absorption process. They are directly linked in the form in which they are absorbed to organic compounds such as sugars, fats, oils and sterol, starches and special kinds of proteins. Phosphorus is an integral part of the plant processes that keep it functioning. Some of these functions are:

a. Phosphorus Carriers

Two compounds, known as ADP and ATP, act as phosphorus carriers to transfer phosphates from one carbohydrate to another during their conversion processes.

b. Hydrogen Donor

A phosphate protein, DPN, supplies hydrogen in some of the biological processes that transform complex carbohydrates into various fats, fatty acids, oils and waxes and sterols.

c. Structural Component

Phospholipids are a very complex group of fatty compounds which contain phosphorus. These control movements of compounds in and out of cells as well as influencing cell contents.

d. Transfer of Heredity

Chromosomes and genes are contained in the nucleus of a plant cell. These determine how a plant grows, what it looks like, how long it grows, the type of flower or fruit it produces, etc. Genes are part of the chromosomes and are composed of nucleoproteins, a type of protein. Other kinds of nucleoproteins are found in the cell fluid, cytoplasm, which is diffferent from the cell nucleus. Phosphorus is in proteins.

e. Release of Energy

All living cells must respire/breathe. Respiration means the use of complex carbohydrates to obtain the energy required for growth and other life processes. When ADP and ATP transfer phosphates to carbohydrates, the links (bonds) holding them together are actually energy warehouses. When bonds are transferred or broken, there is always a release of energy.

h. Phosphorus Deficiency in Plants

Some deficiencies are:

a. Reduced Root Development

Phosphorus is involved in cell division and enlargement as well as the storage and release of energy. A deficiency will stunt root growth resulting in smaller plants.

b. Discoloration

Complex carbohydrates and proteins cannot be formed at normal rates in phosphorus-deficient plants. Thus, such plants tend to accumulate sugars. The plant has a built-in self-protection mechanism which transforms the surplus sugar into pigments that turn the plants red or purple.

c. Maturity

Phosphorus-deficient plants are almost always late maturing. Also, the seeds formed may be shrivelled, light in weight and may have a smaller percentage of yield.

d. Older Leaves Affected First

As the deficiency becomes more severe, the symptoms become more noticeable as they progress from the oldest leaves toward the youngest leaves and fruit buds.

i. Excess Phosphorus in Plants

The roles of nitrogen and phosphorus in plant growth are closely related in many ways. Nitrogen tends to accumulate in plant tissue and does not do a complete job when there is a deficiency in phosphorus. Excessive phosphorus in the soil can depress nitrogen absorption and insufficient nitrogen in the soil can decrease the rate of phosphate-ion absorption. These are just a few examples of the influence of nutrients upon one another. All the various life processes that go on inside the plant are dependent on the supply of plant nutrients and these are dependent on one another. There is no substitute for a balanced nutrition.

★ ★ ★

3. POTASSIUM

Potassium (K) is the third of the three primary nutrient elements—nitrogen, phosphorus, potassium. Except for nitrogen and the secondary nutrient calcium, plants remove more potassium from the soil than any other nutrient. Potassium, often expressed as its compound potash, (K_2O), is essential to the life processes of all plants. It is essential in the manufacture and movement of sugars and starches within the plant and to normal growth through the division of plant cells. Potassium hastens maturity and seed production and aids in root development.

Potassium exists in the soil in several forms. One form is soluble in water while the other forms are insoluble and are unavailable to the plants. Plants can draw upon only 1% of the total supply of potassium in the soil. This supply, known as exchangeable potassium, may be provided by both minerals in soil particles and organic matter. In its exchangeable form, potassium is not soluble and thus is not free to move in soil solution *until* it undergoes a slow weathering process which releases ions into the solution. However, the roots of plants can assimilate exchangeable potassium from the soil or humus through direct contact without it actually ever entering the soil solution.

a. Forms of Potassium in the soil

The ability of a soil to hold nutrients through electrical attraction (positive and negative charges) is determined by the number of negative charges contained by the soil particles or collides. Two factors are important to this nutrient-holding-capacity: the type of clay material and the amount of clay material in that soil. While the type and amount of organic matter in the soil is also important, it is omitted from the following discussion for simplicity.

Clay soils typically found in midwestern and western areas have a much higher nutrient-holding capacity (positive ion exchange capacity) than do the clay soils typically found in southern states. Certain types of clay are the main sources of available potassium. Other soils have less clay or clay with a smaller storage capacity so much of their reserve of potassium has been used up.

In many soils the total amount of potassium per acre ranges from 20,000 to 40,000 pounds for each 6 inch layer of soil, Even so, many of these soils cannot supply enough available potassium to plants. More than 95% of the potassium may be locked up so tightly in the soil minerals that not enough becomes usable each year for the plants. Most of the potassium in soils is in the form of finely ground, weathered rock. It becomes available very slowly over thousands of years through chemical weathering of the minerals.

b. Available Forms of Potassium

There are three forms of potassium in the soil that are available to plants or become available in a few months or years. These are:

    a. the store-house of non-exchangeable potassium between clay particles,
    b. the exchangeable form on the outside of clay particles
    c. the water soluble form in soil solution.

The store-house form is called non-exchangeable or fixed potassium which is held between the layers or plates that make up the clay particles. The potassium stored near the outside of the clay particles can easily move out to become available either as exchangeable potassium or as water-soluble potassium. The exchangeable form is lightly attached to the surfaces of the clay and humus in the soil. It is readily available to crops and makes up most of the soil potassium measured by a potassium test. The other K form measured by a test is in the soil solution. The plant roots contact only a part of the total amount of potassium stored in the soil. The amount removed by plant roots from the soil solution is continually replaced by the exchangeable potassium.

c. Mobility in Soil

Potassium movement in the soil varies with soil types. In most agriculturally important soils, potassium moves to a very limited extent. Potassium can be leached from sandy soils and from soils with a very low exchange capacity, so sandy soils may need annual applications. But where normal rates of potassium-bearing materials are applied, leaching losses of potassium are extremely small under most conditions and negligible or non-existent in most midwestern and western states. Annual application of potassium-bearing materials is preferred to large quantities intended to last for years as it avoids losses by leaching or fixation.

Plant nutrients exist in the soil solution as charged particles called ions. They carry either a positive charge or a negative charge. Nature dictates that electrical neutrality must be maintained at all times. Therefore, a balance always exists between positively and negatively charged particles. This means that when a positively charged nutrient ion, such as potassium, moves in the soil, it must be accompanied by a negatively charged ion, such as chloride, sulfate, phosphate, nitrate, etc. Since these negatively charged nutrients move in the soil at different rates, they govern to some extent the rate of movement of the positively charged nutrients like potassium. Of the three major nutrients (nitrogen, phosphorus, potassium), only potassium has a positive charge under all conditions. Since the soil particles are negatively charged, potassium is readily attracted to and held by these soil particles. This limits the extent of its movement in the soil.

d. Potassium in Acid Soils

In acid soils, a large percentage of the positive ion exchange capacity is taken up by aluminum ions which are tightly held and are replaced with difficulty by potassium ions when they are added to the soil. When acid soils are limed, much of the exchangeable aluminum is replaced by calcium and/or magnesium. This exchange increases the soil pH, increases the positive ion exchange and increases the amount of potassium which can be held. Maintaining a favorable soil pH not only increases crop yields, but also reduces the potential loss of valuable plant nutrients, including potassium.

e. Potassium Deficiencies

The potassium concentration range in a normal leaf is from one to four percent. Potassium deficiency can generally be noted by shortened internodes, loss of the dark green leaf color and in the case of a severe deficiency a burning along the edge of the lower leaves, lodging among small grains or stalk breakage in corn and sorghum.

Many internal growth cycles in the plant are usually disrupted before potassium deficiencies are noted externally. Among them are a reduction in

a. photosynthesis causing slower plant growth,

b. carbohydrate metabolism leading to smaller plants, few grains and/or weaker stalks,

c. protein metabolism which means less protein in forage crops,

d. plant resistance to diseases, so plants are more susceptible to fungus, mildew, bacterial wilt, etc.,

and e. plant water relationship. Adequate potassium maintains higher moisture levels.

f. Effects on Fruits and Vegetables

Another function of potassium which results from carbohydrate, protein and organic acid production, is the quality of fruits and vegetables. Potassium has been known to increase such factors as fruit size, fruit color and to improve keeping quality. Quality increases are usually brought about over the same potassium range that is required for optimum yields.

★ ★ ★

4. SULFUR

Sulfur, one of the secondary nutrient elements, plays a dual role in the nutrient and soil functions required for plant growth. As a nutrient it is essential to the life of the plant and as a soil additive, it has the unusual ability to bring about physical and chemical changes in the soil. Sulfur is required for synthesis of sulfur-containing amino acid compounds. A deficiency of sulfur will cause plants to exhibit characteristics similar to those with a nitrogen deficiency.

Sulfur is generally added to the soil to perform one or a combination of these functions:

a. supply sulfur as a nutrient,

b. lower the pH of naturally alkaline or over-limed soils,

c. increase the availability of other plant nutrients such as phosphorus, iron and manganese,

d. improve the physical condition of the soil, and

e. alkali and saline-alkali soils.

Plants usually absorb sulfur as the *Sulphate* ($SO_4^=$) ion, However, the sulphate ion is generally not retained in the soil very long. The sulphates are water soluble, tend to move with soil water and are readily leached from the soil under high rainfall conditions. This is particularly true of sandy soils.

a. Forms of Sulfur in Soil

Sulfur can occur naturally in the soil or can be added in several different ways. Sulfur generally occurs in the organic matter of soils at a ratio of 10:1 with nitrogen i.e., 10 parts nitrogen to one part sulfur. Organic compounds of sulfur exist as organically bound sulphates in the soil and in other compounds.

b. Farmyard manure

This is an excellent source of sulfur and in the past, farmyard manure has been one of the primary contributors of sulfur to the soil. Another source is the sulfur dioxide (SO_2) emitted into the air by the industrial burning of sulfur-containing fuels such as coal and oil. The sulfur dioxide is then brought to earth by rainfall. However, both of these sources are rapidly disappearing from use. People are hesitant to use farmyard manure and yet it is one of the best all-round organic soil additives available. The restrictions by the Environmental Protection Agency on free emission of sulfur from smoke stacks and, more importantly, the concentration of such emissions in urban industrial areas, have decreased the amount of sulfur available free to rural areas from rainfall.

Elemental sulfur (S) can be added to the soil in a finely ground form to increase the nutrient base. Elemental sulfur does not dissolve in water, nor can it be mixed with water without the use of special materials.

c. Transformation of Sulfur in Soil

Elemental sulfur is oxidized to sulfuric acid (H_2SO_4) by soil bacteria. Other factors such as moisture and temperature being equal, the rate of plant-nutrient sulfur availability from these sources is determined by the fineness of the elemental sulfur. With finely ground sulfur, oxidation is fairly rapid in most moist, warm well-aerated soils.

d. Effect of Sulfur on pH

One purpose of adding sulfur to the soil is to bring the pH down to an ideal level for plant growth. Soil micro-organisms convert sulfur into sulfuric acid (H_2SO_4). The sulphate particle, the $SO_4^=$ ion, of the acid is the plant nutrient, while the acid portion, the H_2^+ ion, acts as the soil amendment. It is this hydrogen ion that makes the soil more acid. Various soils need varying amounts of sulfur to affect a pH change.

In addition to creating a more desirable environment for plant growth, lowering high soil pH values tends to increase the availability of phosphorus and certain secondary and trace elements. Plants usually affected by these alkaline-induced deficiencies include soybeans, other beans to a lesser extent, wheat, oats, citrus fruits, ornamental plants and certain lawn grasses.

In soils with excess sodium or sodic soils, the desirable calcium (Ca^{++} ions) or magnesium (Mg^{++} ions) of the soil particles are replaced by the undesirable sodium (Na^+ ions). There may also be an excessive accumulation of soluble salts in the soil. The action of sulfur and sulfur compounds on these soils is to supply calcium or release insoluble calcium already present. This calcium replaces the absorbed sodium on the clay particles. The sodium then combines with the sulfate ion to form soluble sodium sulfate which can be removed by leaching.

When the soil particles contain a balance level of calcium ions (65 to 75% of the exchangeable capacity), the particles join together in groups to form larger soil particles called *"aggregates"*. The formation of aggregates permits water to percolate through the soil, oxygen to enter and thus for plants to grow better. The percolating water carries the displaced sodium and other salts downward and, if soil drainage is good, the sodium and salts are flushed out of the soil. If the problem soils do not contain free lime (calcium carbonate), then gypsum is usually the preferred source of material for reclamation.

e. Sulfur in Plants

Plants cannot absorb elemental sulfur, so it must be changed to a soluble oxidized sulfate ($SO_4^=$) form in order to function as a nutrient. Plant requirements for sulfur vary. Cabbage, kale, cauliflower, turnips, radishes, onions and asparagus have very high sulfur requirements. A 15-ton crop of cabbage, for example, takes up about 40 pounds of sulfur per acre while a 15-ton crop of onions takes up about 30 pounds per acre. Plants with medium sulfur requirements are alfalfa and sugar beets, while plants with low sulfur requirements are small grains, grasses and corn.

f. Sulfur Deficiency in Plants

Plants suffering from sulfur deficiency usually show characteristic, but not always readily distinguishable, symptoms. These systems are so similar to those of a nitrogen deficiency that they may be mistaken for that lack. Typical sulfur deficiency symptoms are

    a. plants are small, spindly and have short stocks,

    b. plant growth rate is usually retarded and maturity is often delayed,

    c. young leaves on most plants will be light green to yellowish with even lighter-colored veins. On plants such as citrus and cotton, some of the older leaves will be affected first. Spotting of leaves may also occur on vegetables such as potatoes.

    d. formation of nodules on legumes is often retarded or reduced,

    and e. in many cases, fruits do not fully mature or are light green in color.

In many cases, a sulfur deficiency may be caused by a sandy soil that is highly leached and low in organic matter content.

★ ★ ★

5. MAGNESIUM

Magnesium is another one of the three secondary nutrient elements required by plants for healthy growth. Magnesium occurs abundantly in nature; it comprises approximately 1.9% of the earth's crust. Magnesium is the *only* mineral or metallic element which is necessary to activate the enzyme system and the oil content of plants. A deficiency of magnesium in plants can restrict growth and yields.

a. Magnesium Deficiencies in Soil

Magnesium deficiencies generally occur in deep, coarse textured, acid soils with a pH of 5.2 or less. These soils usually have a low positive ion exchange capacity and have had excessive leaching. In soils with a pH of 5.8 or more, a deficiency rarely occurs even though the amount of available magnesium is low. Other factors which can contribute to magnesium deficiencies in the soil are:

a. application of high calcium and low magnesium materials to soils already low in magnesium,

b. addition of large amounts of potassium and ammoniacal-nitrogen bearing material to soils already low in magnesium, and

c. continuously growing crops that require large intakes of magnesium on the same land.

b. Magnesium Deficiency in Plants

The appearance of leaf symptoms is generally the first visual indication of magnesium deficiency. Magnesium is mobile within the plant and is readily translocated within the plant. Deficiency symptoms appearing first on the older leaves are:

a. loss of healthy green color between the veins, followed by chlorosis which may start at the leaf margins or tips and progress inward,

b. leaves may be abnormally thin,

c. leaves tend to be brittle and curve upward, and

d. with severe deficiencies, leaves tend to take on a reddish purple color on tips and edges.

★ ★ ★

6. CALCIUM

Calcium is the third of the secondary nutrient elements. The effects of calcium on the soil are discussed in detail in conjunction with soil pH and with sulfur.

Calcium is necessary to maintain the soil pH at the level required by plants for the assimilation of the minerals in the soil and to make these minerals readily available to plants. Within plants, calcium is associated with activating the plant's enzyme system.

A deficiency of calcium, rarely observed in western soils, can cause a gelatinous sticky substance to occur at the tips of new leaves, causing them to stick together. As noted previously, an alkali soil high in sodium can interfere with the calcium metabolism of plants.

★ ★ ★

7. BORON

Boron is a trace nutrient element which occurs in extremely small quantities in the soil, approximately 20 to 200 parts per million. But in some western soils and in irrigations waters, boron can occur in *toxic* quantities. Boron is involved in activating the plant's enzyme system and influencing plant cell development. Boron is only required in minute quantities for adequate plant development.

A deficiency of boron can occur from leaching in sandy soils low in organic matter and with high boron demanding plants such as alfalfa. A highly alkaline soil of pH 8.0 or more with too much calcium carbonate can chemically render boron unavailable to plants. A deficiency of boron can result in cessation of growth of terminal buds in plants and internal corking or blocking of root crops and deciduous fruits.

★ ★ ★

8. IRON

Iron is one of the more important of the trace elements necessary for plant growth. Iron is essential to plant nutrition since it is the activating element in several enzyme systems. An enzyme is a complex organic compound which causes chemical changes. Iron is essential for photosynthesis, for formulation of chlorophyll and for transformation of carbon dioxide into plant energy. Iron is also important in respiration and other oxidation within plants as it is a vital part of the oxygen carrying system. Since iron is immobile in plants, an iron deficiency will first be seen in the new leaves on plants.

There are a number of factors that bring about iron *chlorosis*—yellowing of plant leaves. It is seldom the case that the soil is lacking in iron. Usually there is enough iron present in the soil; however, it is in compounds unavailable to the plants.

A high pH and mineral imbalance are often responsible for this tie-up or unavailability of iron in the soil. The presence of bicarbonate ions and high phosphate levels create deficiencies and phosphates tend to immobilize iron in the roots and younger leaves of plants. An excess of the major metal elements can cause an imbalance in the trace elements.

In western soils, iron chlorosis usually occurs as a result of an excess of calcium carbonate (lime). This deficiency is often referred to as "lime-induced chlorosis". These calcareous soils are alkaline and have a pH well above 7.

In certain parts of Florida, acid sandy soils frequently develop chlorosis as a result of copper toxicity. This may be particularly evident where copper sprays are used for disease control. Similar toxic effects also develop from other metals such as manganese, zinc, molydenum, cobalt and chromium.

Other factors can also interfere with the absorption or utilization of iron by the plant. Excess water or over-irrigation and poor aeration can create unfavorable root environment for the uptake of iron. Soil and climatic conditions such as temperature, light intensity and levels of organic matter as well as viruses, nematodes and other soil organisms can also affect the availability of iron.

★ ★ ★

9. ZINC

Plants only use a very small amount of zinc; however, it is one of the essential nutrients for plant growth. A certain level of available zinc must be maintained at all times for good plant production. Without adequate zinc, the vigor of the plant is reduced, growth is slower, maturity is delayed, yields are reduced and quality is affected. Also, developing fruits can be small and malformed.

Zinc is related to the normal use of carbon within plants as it is need for protein metabolism. It forms part of the enzyme system which regulates plant growth. Zinc is concentrated in plants where growth is greatest: in shoot tips where stems and leaves are formed, at nodes where buds form and leaves grow and in seeds where it is readily available for the young seedling. A vigorously growing plant, therefore, needs more zinc than one which is developing slowly.

a. Zinc Availability

Certain practices such as cultivation and fertilization can affect the availability or uptake of zinc. Some plants such as sugar beets, will deplete the zinc supply in the soil. Soil type, pH and organic matter in the soil can also affect the zinc availability as zinc has a tendency to become fixed or tied-up or non-absorbable by the plants.

The total amount of available or unavailable zinc is low in acid, sandy soils which are subject to leaching. Soils formed from granite-like rock may also be low in zinc. Fine textured soils generally contain more zinc than do coarse textured ones and topsoil contains more zinc than does the subsoil. Calcareous materials such as limestone increases the zinc-fixing.

Zinc is most soluble and available to plants under acid soil conditions where the pH is below 6.0. At pH 6.0, the zinc availability is low and as the pH increases, the availability continues to decrease. Nearly all zinc in the soil becomes fixed at pH 9.0 and is completely unavailable to plants. Soils high in natural phosphates are often low in available zinc, especially if that soil has a high pH value. High fertilization will increase zinc deficiency since the total plant growth is stimulated, thereby increasing the zinc requirements beyond the available supply.

When elemental zinc forms are added to the soil, zinc ions are rapidly tied-up and held tightly to soil particles. Zinc then becomes concentrated in the topsoil and is usually kept from moving into the subsoil. For this reason, a zinc deficiency is likely to occur in a number of spots in recently levelled fields or where deep cuts or severe erosion has exposed zinc deficient subsoil. Deep rooted plants suffer most as zinc is not carried into the subsoil.

A high organic matter content in the soil may also contribute to a zinc deficiency. Some investigators consider that organic matter is a major factor in fixing zinc and in making it unavailable as micro-organisms in the soil play a role in fixing zinc.

★ ★ ★

10. MANGANESE

Manganese is usually more abundant in plants than any of the other trace elements. Since plants can use manganese over and over, only small amounts are required. However, shortages can develop and where there is a severe shortage of manganese, there will be defoliation, loss of vigor and lower yields.

Manganese can be a part of the enzymes but usually functions with the enzyme systems of plants. It acts as an activator and a lack of it can limit or stop plant growth. Also, manganese is important in photosynthesis. The food making process is slowed down when there is a manganese deficiency. Leaves often become light green, indicating that a lack of manganese is affecting chlorophyll production.

As with other trace elements, there is normally enough manganese in the soil. Organic soils, however, usually contain less manganese than do mineral soils. Certain soil conditions contribute to the availability of manganese in the soil.

a. Manganese Availability

Soil pH is probably the most important factor affecting the availability of manganese. In acid soils with a pH of 4.0, manganese is extremely soluble and available to plants. In fact, manganese toxicity usually happens at pH 4.0. As the pH is raised, the solubility and availability of manganese is decreased. At pH 6.5 and above, manganese is not readily available to plants. As the pH increases, manganese is changed into insoluble compounds in the soil. Also, microbial activity increases as the pH increases and soil microbes make manganese less available. Organic matter can change or reduce the less soluble manganese compounds into more available forms.

Manganese deficiency is more likely to be a problem in spring time when it is cool and moist. When the temperature is low, manganese is less available and when light intensity is low, manganese is absorbed slowly by the plants. Manganese is less available if soil moisture is maintained at an optimum level for a long period of time rather than the normal occasional drying of the soil.

★ ★ ★

11. COPPER

Although there is much to be learned about how copper functions in plants, it is accepted that copper is an essential trace element. It is a vital part of several enzyme systems in plants. Copper acts as regulator for some of the life processes in plants and also functions in the formation of chlorophyll.

Where copper is deficient, yield and quality are adversely affected. In plants grown with adequate copper, yields are increased and the flavor of fruit and vegetables is improved. Copper increases the sugar content of citrus and it intensifies the color of such crops as carrots, spinach, wheat and apples.

Copper, like the other trace elements, can be deficient in the soil, or it may become fixed or tied up in a form unusable by the plants. Copper leaches easily in deep sandy soils low in organic matter. However, copper moves very little in clay soils and it remains fairly available to plants.

Soils high in organic matter hold copper in a form unavailable to plants. Research has shown that copper remains in the cultivated soil layer as long as five years after application. As soils increase in organic matter, the copper is more tightly tied-up in the soil and cannot be absorbed by plant roots. Copper generally becomes less available to plants as the pH goes up. This is obvious in sandy soils containing very little organic matter, the availability of copper is more closely associated with the organic content than with the soil pH.

A complex relationship exists between the trace elements in the soil. Under certain conditions the uptake of copper can either be decreased or increased, depending upon the relationship of copper with micronutrients such as aluminum, zinc, iron and manganese.

★ ★ ★

12. MOLYBDENUM

Molybdenum is a trace element which can be toxic in amounts over the minute quantities normally contained in soils which is approximately 2 parts per million. Molybdenum is used in the soil specifically for the activation of enzymes. Additionally, certain micro-organisms require molybdenum to fix nitrogen in the soil.

A deficiency of molybdenum can cause legumes to turn pale and yellow and cause interveinal chlorosis.

★ ★ ★

13. CHLORINE

Chlorine is a trace element which is almost never deficient in soils or in plants. Sources of chlorine are sea water, chlorinated city water and sewage water. In many areas, especially coastal ones, a chloride build-up can occur which can be toxic to plants.

★ ★ ★

D. CHELATION

Chelation is a term derived from the Latin word, "chela", which means "pincerlike organ or claw". Chemically, a chelate is a compound formed by combining a metal with a chelating agent. The chelating agent is an organic compound that can act like a claw. It grasps the metal ion and holds it so it can not readily enter into other reactions with the soil.

When a metal is chelated, it is kept in a readily available form until plants want to absorb it. This means the chelated metal is free to move in soil solution within the soil and into the root zone. The chelating agent keeps the metal ions more or less isolated and keeps them from becoming fixed in the soil. Chelation is an important process used to provide plants with the required trace elements.

Although chelation can be chemically accomplished by man, it occurs naturally in soils with a high humus content. In the soil, chelators are substances produced by humus coming in contact with minerals in the soil or with rocks or rock particles. This chelating action by humus makes iron, copper and other trace metals available to plants. These chelated compounds are then continually released as nutrients in the soil as the plants require them.

E. SIGNS OF NUTRIENT DEFICIENCY SUMMARY

Specific nutrient deficiencies have already been discussed, as well as detrimental results of excesses of certain nutrients.

If the soil contains plenty of organic matter or lots of well-made compost, plus some form of additional nitrogen-- bearing material and earthworms, *no other additives* should be necessary to provide plants with the nutrients required for good healthy growth. This is true of most soils in many parts of the United States. Very few soils are deficient in phosphorus or potassium and any such lack can be corrected by adding bone meal or kelp meal to the soil or compost.

The biggest culprit in nutrient deficiency is an incorrect soil pH. This may affect plant growth in various ways at both extremes of the pH scale. Soil pH affects the microbial activity necessary to release or convert nutrients into usable forms for plants as well as plants' ability to assimilate or use those nutrients that are present.

When soil becomes alkaline, various minerals such as iron, manganese and copper are fixed in chemical compounds and are unavailable to plant roots. Plants may show signs of iron deficiency, or yellowing of their leaves—chlorosis. In such cases, the new leaves develop a light yellow color which shows up first between the veins and then spreads to the veins. A deficiency of manganese is similar, where a chlorosis also shows up between the veins on new leaves. But with a manganese deficiency, this yellowing may spread to old leaves. While the chlorotic areas may turn brown or transparent, the veins usually remain green, even with severe deficiencies. In this case, the minerals are present but not available to the plant.

It should be noted that the symptoms of nutrient deficiencies can easily be confused by the layman with similar symptoms caused by plant disease and air pollution. Often many tests and patient observation are needed to pinpoint the cause of poor plant growth or low yields.

★ ★ ★

CHAPTER FIFTEEN

ORGANIC MATTER IN THE SOIL

A. ORGANIC MATTER FOR SOIL ENRICHMENT

Soil scientists now tell us that Grandma wasn't so crazy after all. Remember how she saved all fruit, vegetable, grain and even meat table scraps for the garden compost pile? We thought she was a little strange when she added crushed egg shells and old tea leaves to the mess. But table 15-2 shows what plant nutrients were available in that mix. Once rotted or partially decomposed by micro-organisms, Grandma's compost was a valuable additive to the family's garden. This chapter gives scientific explanations and analyses of organic matter in the soil.

Organic matter was defined as pertaining to or derived from, living organisms containing compounds of carbon. It is a very broad term which may be applied to all plant and animal matter, dead or alive. Organic matter refers to both those plants and animals that you can see and to those microscopic plants and animals that you cannot see in the soil. For agricultural purposes, organic matter refers to the remains of all kinds of plants, animals and micro-organisms in various stages of decay.

Several other terms used in classifying organic matter are:

a.) *Green Matter*—refers to plant material only, may be applied to fresh or dead plant matter such as grass clippings, weeds, green or cured hay, sawdust or dry leaves.

b.) *Leaf Mold*—refers to leaves which have decomposed into a form of compost.

c.) *Organic residue*—refers to many different decomposable organic materials. Typical for such materials are garbage, cannery wastes, wood pulp, refuse from textile and leather industries, apple pomace (skins) of cider mills, cottonmote, and many other waste products. Most of these products are unused today. Millions or billions of dollars are spent annually by industry to dispose of these valuable residues. Using the right techniques, these waste materials can be turned into a valuable source of low-cost plant nutrients and provide industry with another source of revenue.

The term *organic matter* has been used throughout the preceding chapter with the emphasis being on the mechanical or structural improvement brought about by adding such material to the soil. Only slight mention has been made of the nutrient value of organic matter and of the fact that organic matter is the *life blood* of the soil micro-organisms. When once living organic materials are added to the soil, they improve the soil's structure and also set in motion the decay of the organic matter by the action of the micro-organisms. During decomposition, many organic materials yield valuable nutrients which are essential for plant growth. Thus organic matter has a dual relationship with the soil. The first one is structural or physical improvement and the second is chemical or nutrient improvement.

The rate of decay and the resultant products of that decay depend upon the original composition of the organic matter. The soil micro-organisms are constantly at work breaking down the complex compounds in the decaying organic matter into available plant nutrients. When organic matter is first added to the soil, a series of decay actions is initiated, and at each stage of decay, different forms of micro-organisms become involved with the process. Depending on environmental conditions of temperature, water and soil aeration, this microbial activity is most intense on freshly

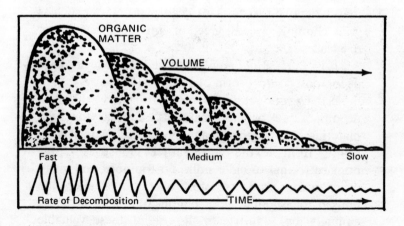

FIGURE 15.1—Rate of organic matter decomposition in soil

added organic material. The microbial activity and population are increased as the soil temperature warms up in spring as long as fresh organic matter has been added to that soil. As the decay action progresses, the organic material obtains a fairly stable, partially decomposed *humus* state. At this point, the rate of decomposition slows down. However, the humus itself will continue to break down into its component elements until its value as a structural conditioner is lost. Therefore, to maintain good soil structure, fresh organic material must continually be added to the soil. Earthworms will greatly speed up the reduction of organic matter into rich, fine-grained humus.

The general change in the chemical form of the organic matter when added to the soil is shown in another manner in Figure 15.1. When the organic matter is first added, the

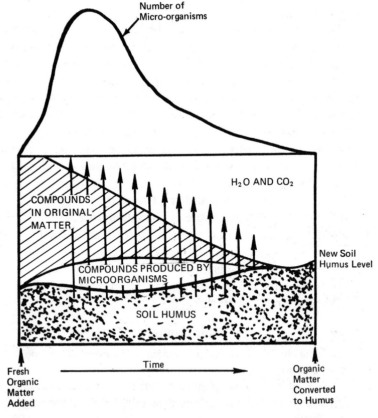

FIGURE 15.2—Effects of decomposition on micro-organisms, synthesized compounds, and humus level

micro-organisms attack the easily decayed compounds such as cellulose and sugar, releasing water (H_2O) and carbon dioxide (CO_2). The micro-organisms also rapidly increase their own numbers as well as the compounds which they synthesize (make). The original soil organic matter is subject to some breakdown as the micro-organism population increases. Water and carbon dioxide are probably the by-products of this breakkdown. When the easily decomposed organic materials are depleted, the micro-organisms start to decline in number. The remaining micro-organisms then start to work on the more resistant materials and compounds in both the original organic matter and on the compounds that were previously synthesized. Over a period of time, the added organic matter is fully decomposed and becomes indistinguishable from the original soil humus.

B. SUMMARY OF ORGANIC MATTER FUNCTIONS IN THE SOIL

1. Maintains the soil in a looser, more porous condition which allows:
   a. plants to develop a more extensive root system,
   b. better soil aeration,
   c. more water to enter the soil at a faster rate,
   d. less leaching of valuable nutrients from the soil,
   e. a larger amount of moisture to be stored in the soil for later use by plants and
   f. less soil compaction and crust formation.
2. Maintains the soil at a more even temperature.
3. Furnishes many nutrients necessary for plant growth.
4. Stimulates micro-organism activity necessary to make the nutrients available for plant growth.
5. Holds available nutrients in soil solution or on soil particles until they are needed by the plants.

Your Compost

Here are some points to remember in maintaining your own compost.

1. Add only decomposable (organic) materials from ground bones to grass cuttings. See Table 15.3 and use your imagination.

2. Composts need moisture. An ideal range is from 40% to 60% moisture in your mix.

3. Composts need aeration. Make yourself "smokestacks" of perforated cornstalks, metal or plastic pipes, or bamboo rods.

4. It is usually wise to add good nitrogen sources.

5. Always try to use the same location in your garden for your compost. Enough microbes, bacteria and fungi will remain in the same place at the lowest levels. Then when you use the top levels on your garden, these will start the decomposing process over.

6. As composts "heat up" in the early stages of decay, you often see a cat or dog asleep atop one on a frosty day, as the center temperature can reach over 150 °F!

7. Your compost needs to reach, at least, 55 °F to be active, but temperatures over 130 °F will kill earthworms. You could add your earthworms after the initial "hot" decomposition is over. Once earthworms are active, there will be no more need to mix and turn your compost.

8. In climates with frost in winter, you can insulate your compost with bales of hay which will gradually become mulch. OR you can dig a cave on the side of a hill for your compost pile. Then its inner temperature will remain high enough to decay all winter.

9. A compost pile should ideally be kept at a pH range between 6.5 to 7.5. That range suits earthworms and most plants.

10. Do not make your compost pile too large. If a great deal of material is available, then build a long ROW of compost. If it is more than 4 feet deep or wide, it is too difficult to aerate and maintain, while one less than 2 to 3 feet high is likely to freeze through in winter.

11. The *best* compost mixtures resemble a Super Club House sandwich. They have alternate layers of dry earth, wet green vegetation and then layers of manure, etc.

12. Materials high in Nitrogen and protein are the best natural *activators* so there is no need for commercial compost starters.

13. A good, final compost product looks like earthworm castings, or dark brown crumbs.

14. Also your local library will have books on organic gardening which will have sections on composting. Good Luck!

C. ORGANIC SOIL ADDITIVES

It has been estimated that a soil of average organic content may lose up to 2.5 tons of organic matter per acre per year. This loss then causes a corresponding degradation of soil structure, decrease in microbial activity and a decrease in plant nutrient content. Thus it is an established fact that organic matter should be continually added to the soil for maximum plant production.

Table 15.3 lists the average analysis of various organic materials which can be added to the soil to provide bulky organic matter as well as plant nutrients. Keep in mind that the analysis shown is an *average* only. One material obtained from one source can vary significantly from the same type of material obtained from another source. When selecting an organic additive, consider the content and benefits derived from that material against the cost and availability as both can vary greatly depending on the area you live in. Sometimes, it is cheaper to use a large amount of one material with certain elements and then add a small amount of another material to obtain any missing elements.

D. ANIMAL MANURES

Animal manures are an excellent source of organic material and most are high in nutrient as well as bulk content. Unless it is absolutely necessary, avoid commercially bagged manures. In most cases, they are expensive and have been chemically treated to kill weed seeds, etc. In most areas, various types of animal manures can be obtained free or for a minimal charge from dairies, feed lots, stables, chicken farms or rabbitries.

The one major problem in using animal manures is some types, especially dairy, steer and horse manures, contain a large percentage of *salts*. This means that these manures should be leached before planting. This can be accomplished by leaching the manure before it is applied to the soil as explained on page 115 of Volume I, or manure can be turned into the soil in the fall or winter to let the rains leach it before spring planting. Even watering the manured soils over the winter can leach away the salts. However, leaching will remove some of the nutrients from the manure; a concentrated additive, such as bone meal or greensand, can be added

TABLE 15.3—Average analysis of organic materials

Material	% N Nitrogen[4]	% Protein[4]	% P_2O_5 Phosphoric Acid[5]	% K_2O Oxide[6] (Potash)	% Organic Matter	Cubic Feet per Ton
Dairy Manure	0.7	4.38	0.30	0.65	30	55
Goat Manure	2.77	17.31	1.78	2.88	60	70
Hog Manure	1.0	6.25	0.75	0.85	30	60
Horse Manure	0.7	4.38	0.34	0.52	60	75
Poultry Manure	1.6	10.00	1.25	0.90	50	50
Rabbit Manure	2.0	12.50	1.33	1.20	50	70
Sheep Manure	2.0	12.50	1.00	2.50	60	70
Steer Manure	2.0	12.50	0.54	1.92	60	70
Poultry Droppings	4.0	25.00	3.20	1.90	74	55
Seaweed (Kelp)	0.2	1.25	0.10	0.60	80	—
Alfalfa Hay	2.5	15.60	0.50	2.10	85	—
Alfalfa Straw	1.5	9.38	0.30	1.50	82	—
Bean Straw	1.2	7.50	0.25	1.25	82	—
Grain Straw	0.6	3.75	0.20	1.10	80	—
Cotton Gin Trash	0.73	4.56	0.18	1.19	80	—
Winery Pomace (Dried)	1-2.0	6.25-12.5	1.50	0.5-1.0	80	—
Castor Pomace	6.0	—	2.5-3.0	0.50	80	—
Olive Pomace	1.2	7.50	0.80	0.50	80	—
Apple Pomace	0.2	1.25	0.02	0.15	—	—
Coffee Grounds (Dried)	2.0	12.50	0.40	0.70	—	—
Tea Grounds	4.0	25.00	0.60	0.40	—	—
Oak Leaves	0.8	5.00	0.40	0.02	—	—
Pine Needles	0.5	3.13	0.10	0.03	—	—
Egg Shells	1.2	7.50	0.40	0.15	—	—
Peanut Hulls	1.5	9.38	1.20	0.80	—	—
Rock Phosphate	0.0	—	25-30	0.00	—	—
Greensand (Galuconite)	—	—	1.50	6.00	—	—
Granite Meal	—	—	—	3-5	—	—
Wood Ashes	0.0	—	1-2	3-7	—	—
Dried Blood	13.0	81.25	1.50	—	80	—
Septic Sludge (digested)	2.0	12.50	3.01	—	50	—
Nitroganic	6.5	40.63	3.40	0.30	80	—
Milorganite	5.0	31.25	2-5	2.00	—	—
Tankage	7.0	43.75	8.60	1.50	80	—
Bat Guano[3]	13.0	81.25	5.00	2.00	30	—
Cottonseed Meal	6.5	40.63	3.00	1.50	80	—
Coca Shell Meal	2.5	15.63	1.50	2.50	—	—
Soybean Meal	7.0	43.75	1.50	2.00	—	—
Steamed Bone Meal[2]	2.82	17.63	14.99	—	—	—
Raw Bone Meal[2]	4.27	26.69	10.93	—	—	—

[1] These are average content figures taken from United States and Canadian Agricultural Journals, 1969.

[2] Average analysis of 22 samples, from *Western Fertilizer Handbook*.

[3] Bat guano varies from 2 to 5 times the average contents depending on conditions in the caves.

[4] Nitrogen is a relatively constant component of protein. Thus, the % of protein can be calculated by multiplying % of N content by 6.25.

[5] The % of P_2O_5 X 0.44 = the % of elemental P.

[6] The % of K_2O X 0.83 = the % of elemental potassium.

to the soil before planting. The quality of the manure and its value to the soil can be improved if it is composted before it is applied to the soil.

Surprisingly, another organic source of nitrogen is urine, human or animal, and it is safe to use on plants and in the garden. Urine contains up to 46% nitrogen and is low in calcium but contains some salts which may be left in the soil. A human generally passes up to 12 pounds of nitrogen per year in urine. Nitrogen is normally used at the rate of 100 to 300 pounds per acre by farmers for high productivity. Therefore, each person passes enough nitrogen to fertilize 3,000 square feet per year. A small amount of lime or gypsum can be added to the soil each year to remove any salts which may build up. The Chinese have farmed the same lands for thousands of years using human manures as their basic fertilizer.

PHOTOGRAPH 15.4—A truckload of manure.
Here is Scott Van Heck. Mr. Van Heck travels with Mr. Gaddie to teach classes on the care of earthworms across the United States.

Soil Tests or Common Sense?

If you are not sure which nutrients are available in the soil and which are lacking, a soil test can be done. However, a professional laboratory test is too expensive unless you are farming on a large scale basis and home tests are generally not accurate. There are four basic steps to ensure nutrient availability.

1. Make a high quality compost and continually apply it to the soil in large amounts. If the compost contains a mixture of grass, household garbage, leaves, weeds, etc., it should provide a balance of nutrients through the decomposing action of soil micro-organisms.

2. Add an additional organic nitrogen source to the compost. Since nitrogen is easily lost and large amounts are used by plants, a deficiency could result without a supplemental source.

3. Add a reasonable amount of earthworms if your soil lacks them.

4. Learn to recognize symptoms of nutrient deficiencies in plants so you can correct them.

In most cases in a soil with a very high organic content and a correct or near neutral pH, plants will receive all of the nutrients they require.

★   ★   ★

CHAPTER SIXTEEN

WORMS AND THE FUTURE

In Chapter 12 to 15, we learned the basic facts about the natural environment of the earthworm. With a better understanding of these facts, we can better appreciate the several roles played by our friends beneath the surface of the ground; how they continually turn, aerate, and structure the soil; and how they transform nutrients contained in decaying plant and animal matter into forms usable by the next generation of plants. We have seen how Man can harness these creatures to help him till his soil, produce more and better crops, consume the waste from cities and towns, perhaps even feed his poultry, pets, or other livestock or himself!

A. HOW CHANGES HAPPEN

All of these are uses that Man can make of the earthworm today, or in the immediate future. What is necessary to make these uses happen on a significant scale is a willingness on the part of those now within the vermiculture industry to seek out one project, and then begin working quietly, patiently, and persistently to bring that project about. Just dreaming about how wonderful it would be if your town started using annelidic consumption to dispose of its refuse will not make that happen. You have to go to the people responsible in your town for refuse disposal, present the program to them, and develop a plan for testing and developing the concept in ways that fit your local situation. If you need technical advice, it can be obtained, with suitable arrangement, from NABF as suggested in Chapter 9.

The same principle applies to whatever use of the earthworm you find most significant. None of it will happen by just dreaming about it and wishing it would be so. If you want to promote the use of earthworms and castings by gardeners in your area, get in touch with the horticulture specialist of your Cooperative Extension Service. Or with the garden editor of your local paper. Propose a planting trial. Or go to the gardening instructor of your local high school adult education department, with the same idea. *The key is action.*

Joining together in Farm Bureau with other vermiculturists is most important. We need to come together to share progress, problems, and opportunities. By forming vermiculture commodity units within County, State, and American Farm Bureau Federations, we provide greater visibility for our industry; and increase the likelihood that governmental and university departments will respond more quickly and favorably to our needs.

B. MORE RESEARCH

Even while we work to utilize more fully our present knowledge about earthworms, there are hints that even more wonderful uses may emerge in years to come. Scientists at the University of Ohio have suggested that earthworms could be used to restore strip mine spoil banks. Scientists at the University of Wisconsin reported to a recent meeting of the American Soil Science Society on the possibility of using earthworms as monitors for heavy metal pollution in soils. Another scientist at Xavier University in New Orleans is examining the possibility that earthworms can also be used to monitor toxic pesticide and herbicide levels in agricultural soil.

In medieval Europe, earthworms were used in folk medicine as a remedy for arthritis. No one now knows whether or not there was any real medical validity to this use of worms, but it is a fact that many other folk and tribal remedies have been found to have genuine justifications. Is there some substance in the earthworm which could if properly prepared, extracted, distilled, provide a cure for this crippling illness? No one knows, but maybe it's worth looking into.

Indeed, we know so little about earthworm biochemistry, that one cannot help but believe that there are a number of significant discoveries which might be made in the course of a thorough examination of this field with modern equipment and techniques. It has been speculated by some scientists that earthworms produce *auxins* (a type of micro-chemical) which act as direct stimulants of plant growth. If the earthworm does produce such substances, and if they could be stabilized or synthesized, what a boon it might be for hydroponic farming!

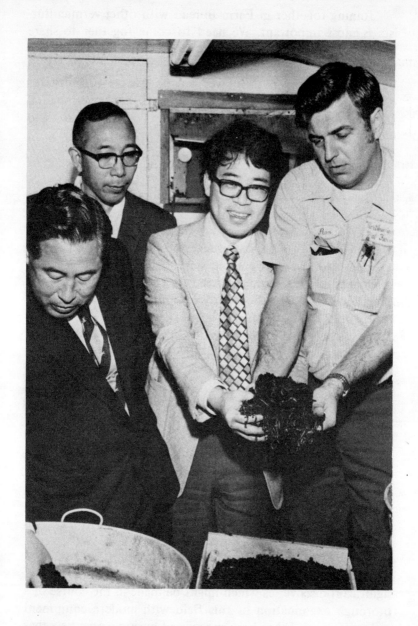

PHOTOGRAPH 16.1—Researchers and worms
Dr. Manabu Nakai and two associates visited Mr. Gaddie at North American Bait Farms in Nov. 1976. Mr. Gaddie will be in Japan, Aug. 1977, to do research with Dr. Nakai.

It has been reported that medical scientists in Japan are much intrigued with the fact that earthworms manufacture *cellulase,* an enzyme which aids in the decomposition of cellulose. Other substances secreted by earthworms may prove valuable to medicine and industry in the future, but only if more research is done.

In the very early part of this century, in the Deep South, thousands of farmers began raising a new kind of crop, one for which few uses were known at the time. Their purpose in raising the crop was to improve their soil for other crops. But they soon found themselves with tons of the new plant, and no market for it. Then a remarkable man went to his laboratory and began to explore all of the possible ways this new crop could be used. In short order he found that it could be eaten several different ways. But equally important were the oils and other substances which could be obtained from this one plant. Today, thanks to that one man, we make more than 200 products from this one plant. The plant was the peanut. George Washington Carver the 2nd, where are you? **The earthworm is waiting.**

★ ★ ★

CHAPTER SEVENTEEN

BOOKS ABOUT WORMS

INTRODUCTION

One of the finest "springboards" to volume earthworm sales, in the experience of many successful growers, is the advance sale of earthworm books and manuals to customers and prospective customers. Any grower or potential grower may obtain from these books, in a few hours of comfortable reading, all of the hard-won knowledge that pioneer earthworm breeders have wrested from many decades of costly "trial-and-error" methods or from a hundred years of tedious scientific experimentation.

All of the earthworm books described in these pages are available for resale, through established growers, at liberal wholesale prices which enable the distributor to make a welcome extra profit on his book sales—enough, in some instances, to pay his entire advertising expense. At the same time, such book sales do much to promote volume earthworm sales.

The books may be purchased for stock at the maximum discount for reshipment by the grower/distributor, or at just a slightly smaller discount (the cost of postage and envelope) for drop-shipment by the publisher. Complete details of this plan, together with samples and prices of book folders, may be had by writing to Bookworm Publishing Company, 1207 South Palmetto, Ontario, Ca. 91761 or your local dealer.

One of the most satisfactory things about book sales is the fact that the sale of one book usually leads to repeat sales of one or more additional books ... sometimes the entire "Earthworm Library," amounting to $20 or more. Most of the books you will sell carry the complete book listings in the back with an order form instructing the reader to buy books from you, the distributor. Average book sales, according to the records of distributors, run around $4 to $5. Attractive six-page book folders for enclosure with your mailings (where, as a rule, they "ride free" so far as postage is concerned) are available to you at cost from Bookworm Publishing Company.

PHOTOGRAPH 17.1 – Earthworm books for extra profits.

Book Postage Rates

Books have a special rate. It is cheaper than the regular parcel post rate. This applies mostly to those instances when you mail several books. Single books which weigh up to 4 ounces would go as third-class mail. When the weight is over that, the package should be marked "Special 4th Class Rate-Book" and mailed at the book rate. This is a flat rate per pound to any city in the United States.

EARTHWORMS FOR ECOLOGY AND PROFIT $4.95
                                        Hardcover $9.95
Volume I, Scientific Earthworm Farming, by Ronald E. Gaddie, Sr., and Donald E. Douglas.

The most comprehensive publication on earthworms and scientific earthworm farming available today. Contains details and information which have not been available to earthworm growers in the past. Reveals the "secrets" that books and growers won't tell; all available markets, planning and installing the earthworm farm, beds and bedding; includes soilless bedding, feeds and feeding methods, how to care for earthworms, how to double the size of earthworms in two weeks. Over 175 pages, completely illustrated. **Available from Bookworm Publishing Company or the dealer from whom you purchased this book.**

EARTHWORMS FOR ECOLOGY AND PROFIT $4.95
                                        Hardcover $9.95
Volume II, Earthworms and the Ecology, by Donald E. Douglas and Ronald E. Gaddie, Sr.

Few people realize the many ways in which the lowly earthworm can help to solve the severe ecological problems which exist on our planet today. This book contains complete and factual information on how the earthworm can be used for ecological purposes such as organic gardening, farming, waste conversion, soil improvement, land reclamation, reforestation, and suburban and lawn beautification. Over 260 pages, completely illustrated. **Available from Bookworm Publishing Company or the dealer from whom you purchased this book.**

LET AN EARTHWORM BE YOUR GARBAGE MAN . $3.50
This report by Home, Farm and Garden Research, Inc., is far more comprehensive than the title indicates. It includes an extensive treatise generously illustrated, on the role of the earthworm in the soil. By Henry Hopp, an eminent authority on earthworms, of the U. S. Department of Agriculture.

DON'T CALL IT DIRT! by Gordon Baker Lloyd $3.95
                                                    Hardcover 8.95
Written by one of America's most experienced and respected garden authors and commentators, this book tells what you need to know to prepare your garden soil for maximum benefits to your plants. Gives instructions on tilling, mulching, adding nutrients, and much more.

DARWIN ON EARTHWORMS by Charles Darwin ... $4.95
Hardcover $9.95
A new edition of Darwin's classic work, *The Formation of Vegetable Mould By The Action of Earthworms,* first published in 1881. A new introduction by James Martin, Professor of Soil Science at the University of California, comments on Darwin's insights in light of recent research.

HARNESSING THE EARTHWORM by Thomas J. Barrett $4.95
Hardcover 9.95
First published in 1947, this is still one of the best books available on organic uses for earthworms. Contains complete instructions on using earthworms for soil conditioning and building for better plants, more beautiful gardens and richer crops.

BIOLOGY OF EARTHWORMS by C. A. Edwards & J. R. Lofty
$6.95
This paperback volume by two leading British researchers is a compilation of more than 600 scientific articles drawn from learned journals all over the world. 283 pgs., including the most complete bibliography on earthworms available today. A must for the serious student or scientist.

WHAT EVERY GARDENER SHOULD KNOW ABOUT EARTHWORMS by Henry Hopp $2.50
Written by a soil scientist formerly with the U.S. Dept. of Agriculture, this little volume details the important effects of earthworms on soil moisture, aeration, nutrients, and crop yields; and details methods for increasing native worm populations in compost pits and gardens.

BEGONIAS FOR BEGINNERS by Elda Haring   Softcover $5.95
Hardcover $10.95
Written by one of America's best known garden experts, a contributor to the New York Times and other national publications, Begonias for Beginners offers both the beginning and the experienced begonia fancier a delightful and thorough look at this fascinating and easy-to-grow plant. Includes over 200 photographs, showing different begonia varieties and culture techniques for both indoor and outdoor growing.

FACTS ABOUT NIGHTCRAWLERS $3.95
Ron Gaddie, calls this "the best book I've seen so far on L. Terrestris." Written by George Sroda, a frequent guest on national television shows, FACTS ABOUT NIGHTCRAWLERS covers many aspects of harvesting, storing, transporting, and selling this most popular of live fishing baits. Fully illustrated with excellent photos and drawings. 111 pages.

BOOK ORDER FORM

This order form is for your convenience in ordering books. All books will be mailed promptly on receipt of price and handling fee. Please order, with your remittance, from the same source through which you purchased this book. If your dealer is out of stock you may order direct from Bookworm Publishing Company. Box 3037, Ontario, California 91761.

Note: Prices subject to change without notice. Date:

Gentlemen: Please send me the books I have listed below, for which I enclose:

☐ Cash   ☐ Check   ☐ Money Order in the amount of $

Quantity Ordered	Description	Unit Price	Total Price
	EARTHWORMS FOR ECOLOGY AND PROFIT, Vol. I, Scientific Earthworm Farming *Soft cover* *Hard Cover*	$4.95 $9.95	
	EARTHWORMS FOR ECOLOGY AND PROFIT, Vol. II, Earthworms and the Ecology *Soft cover* *Hard cover*	$4.95 $9.95	
	LET AN EARTHWORM BE YOUR GARBAGE MAN	$3.50	
	DON'T CALL IT DIRT *Soft cover* *Hard cover*	$3.95 $8.95	
	DARWIN ON EARTHWORMS *Soft cover* *Hard cover*	$4.95 $9.95	
	HARNESSING THE EARTHWORM *Soft cover* *Hard cover*	$4.95 $9.95	
	BIOLOGY OF EARTHWORMS	$6.95	
	WHAT EVERY GARDENER SHOULD KNOW ABOUT EARTHWORMS	$2.50	
	BEGONIAS FOR BEGINNERS *Soft cover* *Hard cover*	$5.95 $10.95	
	FACTS ABOUT NIGHTCRAWLERS	$3.95	
		Subtotal	
		Shipping & Handling	.75c
		State tax (where applicable)	
		TOTAL	

Name _____

Street or RFD _____

City _____ State _____ Zip _____

BOOK ORDER FORM

This order form is for your convenience in ordering books. All books will be mailed promptly on receipt of price and handling fee. Please order, with your remittance, from the same source through which you purchased this book. If your dealer is out of stock you may order direct from Bookworm Publishing Company. Box 3037, Ontario, California 91761.

Note: Prices subject to change without notice. Date:

Gentlemen: Please send me the books I have listed below, for which I enclose:

☐ Cash ☐ Check ☐ Money Order in the amount of $

Quantity Ordered	Description	Unit Price	Total Price
	EARTHWORMS FOR ECOLOGY AND PROFIT, Vol. I, Scientific Earthworm Farming Soft cover Hard Cover	$4.95 $9.95	
	EARTHWORMS FOR ECOLOGY AND PROFIT, Vol. II, Earthworms and the Ecology Soft cover Hard cover	$4.95 $9.95	
	LET AN EARTHWORM BE YOUR GARBAGE MAN	$3.50	
	DON'T CALL IT DIRT Soft cover Hard cover	$3.95 $8.95	
	DARWIN ON EARTHWORMS Soft cover Hard cover	$4.95 $9.95	
	HARNESSING THE EARTHWORM Soft cover Hard cover	$4.95 $9.95	
	BIOLOGY OF EARTHWORMS	$6.95	
	WHAT EVERY GARDENER SHOULD KNOW ABOUT EARTHWORMS	$2.50	
	BEGONIAS FOR BEGINNERS Soft cover Hard cover	$5.95 $10.95	
	FACTS ABOUT NIGHTCRAWLERS	$3.95	
		Subtotal	
		Shipping & Handling	.75c
		State tax (where applicable)	
		TOTAL	

Name_____

Street or RFD_____

City_____ State_____ Zip_____

INDEX OF MAJOR TOPICS

A

Acid soil, 182, 196
African nightcrawler, 12
Agricultural research, 90
Air freight, 160
Airline workers, 160
Alkaline soil, 182, 197
Allolobophora (Eisenia) foetida, 25
Ammino acids, 208
Ammonia, 46
Anerobic digestion, 129
Annelidic consumption, 130, 135, 139
Atlavinyte, O., 17
Auxins, 248

B

Babb, M. R., 136-139
Bacteria, 73
Baeumer, K., 99
Barley, K. P., 112
Barrett, T., 54, 130
Bartlett, A., 116-118
Biodegradable, 128
Biodegradable wastes, 70, 130, 140
Books, 251-254
Boron, 200, 231
Brown-nose angleworm, 25

C

Calcium, 204, 231
Carbon, 200, 213
Carbon/nitrogen ratio, 213-217
Castings, 116-126
Castings—advantages, 116, 124
Castings—composition, 119
Castings—how to produce, 119
Castings—indoor plants, 123
Castings—future, 125
Castings—marketing, 120
Castings—research, 118
Cattle feed, 145-146
Causey, D., 92
Cellulase, 17, 250
Cellulose, 17
Chelation, 236
Chlorine, 206, 236
Clays, 175-176
Combault, A., 2
Compost—produce it, 241-242
Cooking earthworms, 14

Copper, 206, 235
Cossens, G. G., 4
Crump, D. K., 86

D

Darwin, C., iii, 55, 130
Dingwall, A. R., 6
Diplocardia verrucosa, 24
Dixon, R. M. 10
Dreidax, L., 3

E

Earthworms, 21
   adding to garden, 83-85
   aeration & porosity, 12
   aggregation, 61–63
   aquatic ability, 38
   amount food consumed, 59-60
   animal waste, 135-139
   Australia, 114, 162
   biological functions, 27
   biology, 19
   blood circulation, 30
   books, 251-256
   breeding characteristics, 32
   calciferous glands, 41
   Canada, 130
   care, 78-81
   classification, 19
   coelom, 44
   color variations, 40
   commercial production, 159-160
   containers, 159-160
   crop, 42
   digestion, 31, 40
   drainage, 64-67
   effect the soil, 60-77
   environment, 60
   enzymes, 43
   erosion, 67
   excretion, 45
   feed, 34, 155-156
   fertilizers, 110-111
   field crops, 90-91
   food dislikes, 57
   food supply, 57-59
   future, 241-250
   gizzard, 42
   harvesting, 157-159
   head, 30
   home waste, 140

Earthworms. continued:
  human food, 143
  ideal environment, 34
  industrial wastes, 140
  infiltration rates, 65
  intestines, 42
  irrigation, 112-112
  Japan, 130
  life cycle, 35
  life span, 37
  luminescence, 47
  medicine, 248
  migration, 114-115
  mini-farms, 152-155
  moisture, 10-12, 34
  moisture sensitivity, 49
  nephridia, 45
  neutralizing effect, 42, 149
  new books, 166
  nutrients, 15-18
  oxygen consumption, 39
  orchards, 88-90
  organic matter decomposition, 68-70
  pastures, 86-88
  particle breakdown, 60-61
  pet food, 146
  pH sensitivity, 47
  pollution, 127
  populations, 53
  proper pH, 80
  protection, 85
  protein, 144
  questions, 145
  recipes, 147-149
  reproduction, 31
  research, 248-250
  respiration, 37
  segmentation, 29
  senses, 31
  sewage, 140
  soil, 10-18, 179, 183
  soil microbiology, 75-77
  soil monitors—metal pollution, 248
  soil monitors—toxic substances, 248
  soil turnover, 63-64
  soil type, 51-55
  sources, 81-83
  stimulants, 35
  studies, 57-59
  temperature, 34
  temperature sensitivity, 50
  throat, 41
  tillage, 81, 86, 99
  toxicity, 113
  urea, 46
  use, 60-77
  waste disposal, 127
  Worm-gro, 79
Edwards, C. A., 9, 99, 110
Ehlers, Dr. W., 99, 100-110
Elecocytes, 45
energy, 128-130
Environmental protection agency, 135
Eudrilus eugenie, 25

F

Federal government, 127
Field crops, 90-93
Field or garden worm, 24
Fosgate, O. T., 136-139

G

Gaddie, R. E., 147, 249
Gallup, C., 3
Gates, C. E., 93, 114
Gerard, B. M., 92
Gish, C. D., 114
Gore, F. L., 113
Government agencies, 160-161
Grant, W. G., 92
Green matter, 288

H

Hamblyn, C. J., 6
Harvester, D. & T., 157-158
Heath, G. W., 16
Herbicides, 112-114
Helodrilus caliginosus, 24
Helodrilus chloroticus, 24
Helodrilus foetidus, 24
Hinckley, F., 111-112
Home ecology box, 140-141
Hopp, Dr. H., 4, 12, 52
Howell, P., 147
Humus, 240
"Hybrid" earthworms, 26
Hydrogen, 200, 206

I

Inoculating machine, 162-166
Iron, 205, 231-232

J

Jacobson, G. M., 73, 118
Japanese beetle, 160-161

K

Kahsnitz, H. G., 3
Kleinig, C. R., 112
Kuginyte, Z., 115

L

Lakhani, K. H., 17
Laverack, M. S., 144
Leaf mold, 238
Loams, 177
Lofty, J. R., 2, 3, 17

Lumbricus rubellus, 24
Lumbricus terrestris, 23
Lunt, H. A., 73, 118

M

Macro nutrients, 15
Manures, 58, 78
Manure–animal, 243-245
Manure worm, 24
Marshall, V. G., 7
Martin, G. A., 161-166
Micro nutrients, 201
Micro-organisms, 207, 211
Mills, P. A., 112
Mini-farms, 152-155
Molybdenum, 206, 236
Magnesium, 205, 234

N

Native nightcrawler, 23
Nielson, R. L., 4
Nitrification, 211
Nitrogen, 201, 206-217
Nitrogen cycle, 209
Nitrogen deficiency, 212-213
Noble, J. C., 112
Nonbiodegradable, 128

O

Ontario experiment, 132-135
Orchards, 88-90
Organic farming, 1, 116
Organic matter, 238-241
Organic residue, 238
Organic soil additives, 243
Oxygen, 200

P

Pastures, 86-88
Pasture soils, 12
Pest control, 160
Pesticides, 81
Peterson, A. E., 10
pH, 188
pH controlling, 195-198
pH effects, 189
Phosphate, 192
Phosphorus, 202, 217-223
Phosphorus deficiency, 222
Photosynthesis, 185
Pileckis, S., 115
Plant "food", 185
Plant nutrients, 200
Plant–oxidation, 185
Plant–respiration, 185
Plant–soil pH, 191-192
Poloozny, M., 111
Potassium, 203, 223-226

Potassium deficiency, 226
Proteins in earthworms, 144
Pyrolosis, 129
Pastures & earthworms, 162

R

Raw, F., 17
Red worm, 24
Recipes, 147-149
Recipe contest, 142-143
Research problem, 97
Reynolds, J. W., 93
Ribaudcourt, E., 2
Rixon, A. J., 112
Russell, E. J., 3

S

Sands, 176
Schwert, D., 93
Shipping containers, 160
Silt, 176
Slater, E. S., 3
Soil, 167-170
Soil–aeration, 181
Soil–climate, 174
Soil color, 172-173
Soil–compost, 169-170
Soil components, 168
Soil fertility, 183
Soil formation, 170-172
Soil improving, 183
Soil moisture, 180
Soil–organic matter, 168
Soil pH, 182
Soil structure, 178-179
Soil temperature, 181
Soil tests, 184
Soil types, 173, 175-178
Staples, E., 124-125
Staples, W., 124-125
Stockdill, S. M. J., 4, 6, 11, 162
Subsoil, 172
Sulfur, 202, 226-229
Sulfur deficiency, 229
Sulphate, 196
Superphosphate, 218

T

Taff, Dr. D., 71
Taylor, R. L., 150
Taxonomic training, 93-94
Test planting, 94-95
Testing soil pH, 194-195
Thompson, A. R., 113
Tomlin, A. D., 113
Trace elements, 187, 193
Tran-Vinh-An, 77
Treblephosphate, 218

Topsoil, 172
Toxicity, 113

U

Uhlen, G., 6

V

Vail, V. A., 95
Vanagas, J., 18
Van Rhee, J. A., 6
Vermiculture, 95, 151
Vermiculturists, iv
Vermiculturist, training, 93, 96

W

Washington, C., 250
Water, R. A. S., 4
Waste disposal, 127-141
White, G. H., 116
Wildfowl diets, 144
Wormcasts, 116-126
Worm-Gro dealers, 156

Z

Zinc, 205, 232-234
Zrazhevskii, A. I., 6